GCSE STATISTICS

AQA

ully supports the AQA GCSE Specification for 2009 onwards

Practice Book

Delivering the AQA specification

Greg Byrd • Fiona Mapp • Claire Powis • Bob Wordsworth

CONTENTS

Probability

INTRODUCTION

Welcome to Collins GCSE Statistics AQA Practice Book. This book provides lots of extra Foundation and Higher tier content for homework practice and also helps you revise for your GCSE Statistics examination. Use alongside the Collins GCSE Statistics AQA Student Book to get all the information you need for your course.

The content is divided into six main areas of study. These sections are colour-coded:

Planning a strategy

Data collection

Tabulation

Diagrammatic representation

Data analysis

Probability

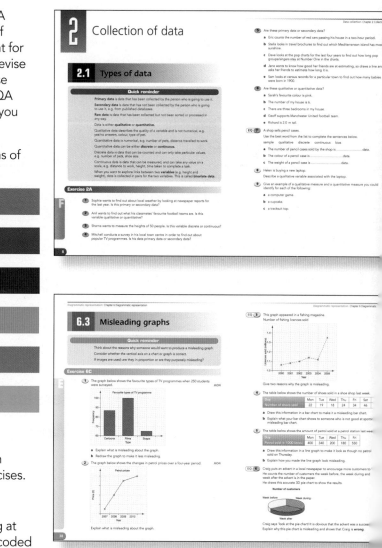

Hints and tips

Remember the key points of a topic with helpful hints and tips at the start of exercises.

Colour-coded grades

Know what target grade you are working at and track your progress with the colour-coded grade panels at the side of the page.

GCSE Statistics Assessment Objectives

Practise the parts of the curriculum where you can gain the most marks (Assessment Objectives AO3 and AO4) with questions that assess whether you can process, analyse and present data appropriately **AO3** an questions that test if you can make deductions and draw conclusions from the data you have collected **AO** There are also plenty of questions that test your basic Statistics skills (AO1 and AO2).

Exam practice

Prepare for your exams with examination-style questions in every topic. Look for (**EQ 1**) in each exercis

Answers

Find all exercise answers on the Collins website at www.collinseducation.com/gcsestatistics

Real-Life Statistics

Find out how Statistics is used in real life with the dedicated section on careers that use data analysis.

1 Planning a strategy

The data handling cycle

Quick reminder

The data handling cycle has four stages:

1 State the hypothesis.

2 Plan the data collection and collect the data.

3 Choose the best ways to process and represent the data (e.g. calculate means etc and draw suitable diagrams).

4 Interpret the data and make conclusions.

You may have to go through the cycle more than once: first of all for a pilot study, perhaps then a further pilot study, and then the actual study.

Exercise 1A

1 Write a hypothesis to be tested in each of these investigations: **AO4**

  a Do boys or girls spend longer doing homework?

  b Does it make any difference if students take a maths exam in the morning or afternoon?

  c Is skiing a more popular holiday now than it was 20 years ago?

  d Does living in the north or south of England make any difference to how long someone lives?

  e Does the rainfall vary more in Scotland or Wales?

2 For each of the investigations in Question 1 above, would you use primary or secondary data? **AO4**

3 For the investigations in Question 1 that you would collect primary data, give an idea of how you would collect it. **AO4**

4 For each of the investigations in Question 1 above, give an idea how you would process the data. What would you calculate? **AO4**

1.2 Planning an investigation

Quick reminder

When you are planning your investigation, it is very important to think carefully about your hypothesis. You must decide what data you will need to test the hypothesis and how you will collect the data. You should also think about what problems you might encounter collecting the data.

As part of your plan, you must think about how you are going to use your data.

Exercise 1B

1 Ying thinks that it is cheaper to buy CDs on the Internet than in high street stores.   AO4

   a State a hypothesis she could use.

   b What data might she collect?

   c What other factors should she consider?

   d What problems might she have collecting the data?

   e How will she use the data?

2 Florence thinks that a person's memory gets worse as they get older.   AO

   a State a hypothesis she could use.

   b What data might she collect?

   c What other factors should she consider?

   d What problems might she have collecting the data?

   e How will she use the data?

3 Roxie has heard the saying 'blondes have more fun' and is thinking she might change her hair colour. Before she does so, she is going to investigate whether the saying could be true.   A

   a State a hypothesis she could use.

   b What data might she collect?

   c What other factors should she consider?

   d What problems might she have collecting the data?

   e How will she use the data?

4 Daniel thinks boys do better at GCSE Maths than girls.   **AO4**

    a State a hypothesis he could use.

    b What data might he collect?

    c What other factors should he consider?

    d What problems might he have collecting the data?

    e How will he use the data?

5 Sam's father thinks that broadsheet newspapers such as *The Times* use longer words than tabloid newspapers such as *The Sun*. Sam is going to investigate this.   **AO4**

    a State a hypothesis he could use.

    b What data might he collect?

    c What other factors should he consider?

    d What problems might he have collecting the data?

    e How will he use the data?

6 Pritesh thinks boys spend more time playing computer games after school than girls do.   **AO4**

    a State a hypothesis she could use.

    b What data might she collect?

    c What other factors should she consider?

    d What problems might she have collecting the data?

    e How should she use the data?

7 Hamish wants to check his hypothesis: **AO4**

'Girls spend less time playing sport than boys do.'

Explain how Hamish should investigate.

2 Collection of data

2.1 Types of data

Quick reminder

Primary data is data that has been collected by the person who is going to use it.

Secondary data is data that has not been collected by the person who is going to use it, e.g. from published databases.

Raw data is data that has been collected but not been sorted or processed in any way.

Data is either **qualitative** or **quantitative.**

Qualitative data describes the quality of a variable and is not numerical, e.g. yes/no answers, colour, type of pet.

Quantitative data is numerical, e.g. number of pets, distance travelled to work.

Quantitative data can be either **discrete** or **continuous.**

Discrete data is data that can be counted and can only take particular values, e.g. number of pets, shoe size.

Continuous data is data that can be measured, and can take any value on a scale, e.g. distance to work, height, time taken to complete a task.

When you want to explore links between two **variables** (e.g. height and weight), data is collected in pairs for the two variables. This is called **bivariate data**.

Exercise 2A

F

 1 Sophie wants to find out about local weather by looking at newspaper reports for the last year. Is this primary or secondary data?

 2 Anil wants to find out what his classmates' favourite football teams are. Is this variable qualitative or quantitative?

 3 Sharna wants to measure the heights of 50 people. Is this variable discrete or continuo

 4 Mitchell conducts a survey in his local town centre in order to find out about popular TV programmes. Is his data primary data or secondary data?

5 Are these primary data or secondary data?

a Eric counts the number of red cars passing his house in a two-hour period.

b Stella looks in travel brochures to find out which Mediterranean island has most sunshine.

c Dave looks at the pop charts for the last four years to find out how long pop groups/singers stay at Number One in the charts.

d Jane wants to know how good her friends are at estimating, so draws a line and asks her friends to estimate how long it is.

e Sam looks at census records for a particular town to find out how many babies were born in 1900.

6 Are these qualitative or quantitative data?

a Sarah's favourite colour is pink.

b The number of my house is 6.

c There are three bedrooms in my house.

d Geoff supports Manchester United football team.

e Richard is 2.0 m tall.

EQ 7 A shop sells pencil cases.

Use the best word from the list to complete the sentences below.

sample   qualitative   discrete   continuous   bias

a The number of pencil cases sold by the shop is ....data.

b The colour of a pencil case is ....data.

c The weight of a pencil case is ....data.

8 Helen is buying a new laptop.

Describe a qualitative variable associated with the laptop.

9 Give an example of a qualitative measure and a quantitative measure you could identify for each of the following:

a a computer game.

b a cupcake.

c a tracksuit top.

10 Are these discrete or continuous data?

   a The weight of a cat.

   b The number of sweets in a bag.

   c Shoe size.

   d The time it takes to boil an egg.

11 Identify the discrete variable and the continuous variable in each of these pairs of variables.

   a The cost of a cup of coffee and the amount of liquid in the cup.

   b The height a boy and his shoe size.

   c The time it takes to travel from home to the station by taxi and the amount it costs.

EQ 12 A doctor's surgery wants to explore its efficiency.

   State whether each of the following variables is qualitative, discrete or continuous.

   a The number of patients seen each day.

   b The time each patient waits to be seen.

13 Rachel is conducting research on trees in her local wood.

   a Copy and complete the table below to describe the sort of data she could collect.

   Use the words: qualitative, quantitative, discrete and continuous.

Data	Type of data
Species of trees	
Number of trees	
Height of trees	
Circumference of tree trunks	
Number of branches	

   b Which two of the data items above could be used as bivariate data?

14 Heather was asked to judge performance in a talent competition. She was asked to rate the contestants as either 1 – excellent, 2 – good or 3 – poor.

   Is this qualitative or quantitative data? Give a reason for your answer.

15 Debbie thinks that a person's age is discrete data. Explain why she is wrong.

16 Choose a variable that could be used to make pairs of bivariate data in each case.

a Car age and _____.

b Height and _____ of people.

c Hours of work and _____.

d Price of house and _____.

17 Give two advantages and two disadvantages of collecting primary data.

EQ 18 A ski resort owner wants to give his guests information about the number of centimetres of snowfall they can expect in January.

a Write down one way he can collect the information if he wants to use primary data.

b Write down one way he can collect the information if he wants to use secondary data.

c Is the data he collects qualitative or quantitative?

EQ 19 Sales staff in a second-hand car showroom have collected information about:

a the most common colours of cars sold in July

b prices of cars sold in July 2010

c number of cars sold in July 2011, compared to July 2010

d makes of cars sold in July

e the average time taken for a member of the sales team to make a sale.

Choose from the terms 'qualitative', 'quantitative', 'discrete', 'continuous' and 'categorical' to describe fully the data in **a**, **b**, **c**, **d** and **e**. More than one term may apply.

2.2 Obtaining data

Quick reminder

Once you have decided which area you are going to research, you need to decide what data you are going to collect and how you are going to collect it.

You can use secondary data e.g. from a database or census, or there are various ways that you can collect primary data:

Experiment

In an experiment, at least one of the variables is controlled – the **independent variable** – and the effect observed is the **dependent variable**.

Example
Sarah carries out an experiment to find out which age group are best at estimating the length of a line.

Age is the independent variable (as she is choosing the age groups), so the estimated length is the dependent variable.

Surveys

Surveys can be done in a variety of ways including through observation, interviews or questionnaires.

Data collection sheets are used when making observations. They help to make collecting the data easier and organise the data as it is collected.

Example
Survey of car colours in a car park:

Colour	Number of Cars
Black	‖‖ ‖‖
White	‖‖ ‖
Grey	‖‖ ‖‖ ‖‖‖
Red	‖‖‖
Blue	‖‖‖‖

Questionnaires are a series of questions that are used to gather data. They ensure that everyone is asked the same questions and that the data is collected in an organised way.

Questions need to be clear, not have any **bias** or express opinions.

Tick boxes can be useful in gathering responses, but in setting these up you need to make sure there are no gaps and no overlaps.

Example
About how many text messages do you send each day?

0 ☐   1–10 ☐   11–20 ☐   21–25 ☐   more than 25 ☐

Data Logging is a mechanical process for collecting data, such as using a rainfall gauge or counting the number of people that pass through a turnstile.

Exercise 2B

 1 Helen wants to find out the number of children in families of pupils in her class. Design a data collection sheet for this data.

 2 Ella is going to spin a coin 100 times and record the result. The coin can land on Heads or Tails.

Design a suitable data collection sheet for Ella.

EQ **3** Jez wants to find out the favourite flavour of crisps in his class.

Design a data collection sheet he could use.

 4 Here is a data collection sheet to be used for a survey about the cost of mobile phones.

Give two criticisms of the data collection sheet.

Cost (£)	iphone	Samsung	Blackberry	Nokia
0–£25				
£25–£50				
£55–£100				
£105–£150				

 5 Alex wants to find out if right-handed people have faster reaction times than left-handed people.

Is the reaction time he measures the dependent variable or the independent variable?

 6 A gym wants to find out when most people use it. Explain how a data logging machine could be used.

 7 A newspaper suggests that the older you are, the more likely you are to watch the news on TV. How would you collect data to investigate this hypothesis?

 8 A publishing company is given a contract to design and market a lifestyle magazine for exercise enthusiasts.

a Explain how they could use both primary and secondary data.

b Explain a possible method of collecting primary data.

9 Maggie thinks that girls are better at spelling than boys. How could she design an experiment to test this.

EQ 10 Abbi wants to investigate the most popular type and memory capacity of personal computer.

a She could use primary or secondary data. How could she get each type of data?

b Design a data collection sheet that Abbi could use to collect primary data.

3 Sampling

3.1 Sampling

> ### Quick reminder
>
> When data is being collected, you will need to take a **sample** of the population being investigated. A **sample frame** is used to identify the population. To avoid **bias**, the size of the sample and the method by which it is selected need to be carefully considered.

Name	Method
(Simple) random sample	Each member of the population is numbered. A random number table or a random number generator on a calculator/computer is used to select members for the sample.
Stratified sample	The population is divided into groups (strata). The same proportion of each group is identified to make up the sample so that the sample mirrors the population as a whole. Members are then selected within each group using a random process.
Systematic sample	A starting point is chosen at random. The data items are then chosen at regular intervals from this point (e.g. every fifth person on a list) depending on the size of the sample required.
Quota sample	A number (quota) of identified groups is interviewed, e.g. 20 men age 20–24 …
Cluster sample	The population can be put into or falls naturally into groups or clusters. A sample of groups is chosen randomly and each member of these groups is included.

Exercise 3A

1 Write the name of the sampling method that is being used in the examples below.

a Andrew needs a sample of 20 people from a numbered list of 100. He generates 20 random numbers and uses those numbered people.

b Emma wants to select a sample of 10 of her class mates. She uses a random number to identify her first person, then takes every third person on the class register until she has her sample.

c A supermarket manager requires a sample of 20 from her workforce of 60 women and 40 men. She randomly selects 12 women and 8 men.

2 Here is an extract from a table of random numbers.

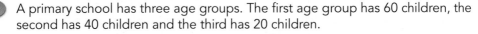

```
48  59  32  38  11  85  84  93  02  29  34  80  94  03  28  57  39  49
52  40  23  21  03  32  72  94  49  49  32  92  02  74  92  24  05  31
47  72  07  94  02  75  69  45  22  84  92  03  39  58  20  05  83  82
82  11  84  74  93  38  37  42  90  31  39  28  59  20  52  43  39  27
```

    a Select 10 random numbers each less than 50. Start at the top left hand corner and work across in pairs from left to right.

    b Select 10 random numbers each less than 50. Start at the top left hand corner and work down in pairs, and form left to right.

EQ 3 A teacher wants to undertake a survey about the number of hours of television that students watch in a week. He considers three possible methods for the survey:

Method 1: Give the survey to the first 40 students seen in a week.

Method 2: Choose 40 students at random.

Method 3: Choose 26 students, picking one whose surname begins with each letter of the alphabet.

    a Give a reason why Method 3 is not suitable.

    b Which of the other two methods for doing the survey will give the most reliable results?

    Give a reason for your choice.

4 A primary school has three age groups. The first age group has 60 children, the second has 40 children and the third has 20 children.

Describe how you would get a sample of 30 children, stratified by age.

5 For each of the following situations:

    i identify the population.

    ii explain why the sample may be biased.

    iii explain a better method to use to choose a sample.

    a Jamilla thinks that girls at her school get more pocket money than boys. There are 300 children at the school: 120 boys and 180 girls. In her survey she asks 30 boys and 30 girls.

    b Kevin wants to find out how far, on average, people in his town travel to work. He asks all the people at his local railway station on a Tuesday morning.

    c To find out attitudes on smoking, an interviewer stopped people in a local shopping centre one weekday morning to ask their views.

6 A local fish and chip shop attempted to estimate the number of people in a certain town who eat fish and chips. One evening they phoned 100 people in the town and asked 'have you eaten fish and chips in the last month?' Forty-seven people said 'yes'. The fish and chip shop concluded that 47% of the town's population eat fish and chips.

Give three criticisms of this method of estimation.

7 A university wants to investigate the use of its canteen. It wants to ask a sample of 50 students in total from three year groups.

Year group	Male	Female
1	600	660
2	420	480
3	480	360

It decides to use a stratified sample.

a Describe the strata it will use.

b Work out the number of males and females in each stratum that will be used.

c Describe how it should choose the individual members of the strata.

8 Henry decides to estimate the number of daisies in the grass on the school playing field. He stands on the playing field and counts the number of daisies within 1 metre of his feet.

a What is the population?

b Why may the sample be biased?

c Describe a better sampling method that he could use.

9 A machine producing corkscrews is believed to produce defective corkscrews at a rate of 10%. The foreman wants to undertake a systematic sample to test this.

a Why might this not be the best method of sampling?

b What would be a better method?

10 The table shows the bookings at a hotel for one month.

Family	Couple	Single Person
93	75	32

The hotel manager wants to send questionnaires to a stratified sample of 30 of these bookings.

Calculate the number of each type of bookings he should include.

EQ 11 The table shows the number of people who use the facilities at a local leisure centre on a particular day.

	Male	Female
Swimming pool	106	15
Gym	180	53
Yoga class	6	24
Spin class	6	10

The leisure centre manager wants to undertake a survey to find the reaction of customers to proposed new opening times.

He decides to take a systematic survey of 20 male gym users.

a Explain how this sample could be collected.

b Give two reasons why this sample would be unrepresentative of the customers as a whole.

c As an alternative, the manager is advised to take a sample, stratified by gender and leisure centre use, of 50 of the 400 customers.

   i Calculate the number of customers at the spin class to be included in the sample.

   ii Calculate the number of female yoga class members to be included in the sample.

12

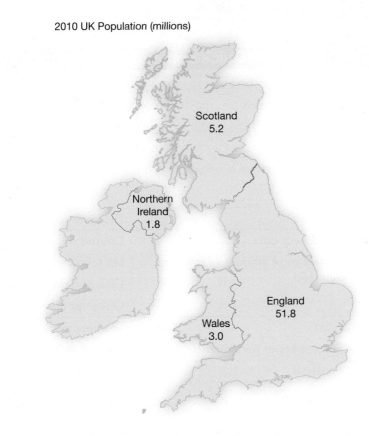

2010 UK Population (millions)

Scotland
5.2

Northern
Ireland
1.8

England
51.8

Wales
3.0

a What was the total population of the UK in 2010?

b Calculate how many people from each country would be in a random stratified sample of 500 000.

4 Conducting a survey or experiment

4.1 Surveys, questionnaires and interviews

Quick reminder

Data for an investigation is often collected by completing a **survey**. This can be done eithe through **observation**, **questionnaire** or an **interview**. Choosing the correct method for the investigation is essential.

A **pilot survey** (trial) is often completed to check that the questions are correct and will giv the required information before the main survey is undertaken. Questions need to be clear with no **bias**.

Survey type	Advantages	Disadvantages
Observation	Easy to collect data Potentially large amount of data can be collected	Limited data can be collected No follow up Lack of detail Observer may need to be trained – cos
Questionnaire	Low cost Large numbers of participants Easy analysis Can be done online	Low response rate from mail surveys Lack of detail
Interview	Personal contact Ability to probe and ask follow-up questions Detailed answers	Time consuming High cost, including training the interviewer People don't like unsolicited telephone interviews

Definitions

A **respondent** is someone who takes part in a survey.

A **pilot survey** is conducted on a small sample to test the design and methods of the surve

A **questionnaire** is a set of questions designed to obtain data.

An **open question** is one that has no suggested answers.

A **closed question** has a set of answers for the respondent to choose from. This may be through, for example, tick boxes or a sliding scale.

Exercise 4A

1 A council included this question in a questionnaire:

'Do you agree that the new one way system is a good idea?'

Give one criticism of this question.

2 The council want to know how many people use each type of recycling bin provided in a local car park. There are bins for glass, tins, paper and plastic.

Design a data collection sheet that will allow you to capture data while at the car park. Remember to include the number of people as well as the type of recycling.

3 Ellie is writing a questionnaire about people's ages. This is one question from her questionnaire.

> How old are you?
>
> ☐ Young   ☐ Middle aged   ☐ Old

a Give one criticism of this question?

b Rewrite the question to make it a good question.

4 Edward is carrying out a survey about rugby teams. This is one question from his survey.

> How often do you watch a rugby match?
>
> ☐ Less than once a week   ☐ Once a week   ☐ Whenever I can

a Give a reason why this is not a good question.

b Rewrite the question to make it a good question.

EQ 5 Krishan receives a questionnaire in the post about a new local leisure centre.

Three of the questions are shown below.

Give **one** criticism of each question.

Question 1:

How often have you exercised in the last 6 months?

Question 2:

How much do you earn each year? Please tick one box.

☐ Less than £10 000   ☐ £10 000 up to £20 000   ☐ More than £20 000

Question 3:

If you have already used our leisure centre, give one reason why you enjoyed using it.

6 Write down **three** of the things you need to consider when writing a question for a questionnaire.

7 **a** Criticise the following questionnaire question:

Approximately how tall are you?

b Rewrite the question to include response boxes.

8 Give **two** advantages and **two** disadvantages of using an interview to gather data.

9 Write a question for a questionnaire designed to find out how much people weigh.

10 A report in a medical journal claims that people weigh more in the morning than in the evening. Design a simple statistical experiment to test this claim.

EQ 11 Max wants to open a shop in the village where he lives.

To find out the views of local people, he delivers a questionnaire to every house in the village.

a The questionnaire includes a closed question about the respondent's age.

i Explain what is mean by a *closed question*.

ii Give one advantage of using a closed question for age.

b Only 14% of the questionnaires are returned to Max.

How might Max have improved the response rate?

c The returned questionnaires showed that some of his questions had been badly worded.

What should Max have done before he delivered his questionnaire to avoid this problem?

d One of Max's questions was:

'How often do you go shopping?'

Give **two** criticisms of this question.

12 A large sports company with 170 shops wants to obtain information about sales.

They decide to send out a questionnaire to all shops, but first carry out a pilot survey.

What are the advantages of conducting a pilot survey?

13 You need to carry out a survey to find out how much money people will spend on a holic

a Give one reason why you might choose to carry out a personal interview rather than a postal survey.

b Give one reason why you might not choose to conduct a personal survey.

c Give one advantage and one disadvantage of conducting an online survey about holidays.

14 Forty white van drivers were asked to complete this questionnaire.

> Throw a coin.
>
> If it shows a HEAD, tick the YES box below.
>
> If it shows a TAIL, answer the question 'DO YOU EVER SPEED WHEN DRIVING FOR WORK?'
>
> ☐ Yes   ☐ No

a How many HEADS would you expect?

b If all the drivers speed, how many forms will have the 'YES' box ticked?

c If no drivers speed, how many forms would you expect to have the 'YES' box ticked?

d When the forms are returned, 32 have the 'YES' box ticked. Estimate the number of drivers who speed when driving for work.

e What is the advantage of using a questionnaire like this?

4.2 Census data

Quick reminder

What's the difference between a census and a survey?

A census collects data from everyone or everything in a population, whereas a survey collects data from a sample of the population as a whole.

	Advantages	Disadvantages
Census	Takes the whole population into account Accurate data Unbiased (as all asked)	Expensive Time consuming Can be difficult to make sure you have responses from the whole population Large quantity of data produced
Survey	Cheaper Less time consuming A more manageable amount of data produced	Not completely representative of whole population Sampling method may inadvertently introduce bias

Exercise 4B

1 An estate agent wants to get information about house prices in the town where he w

  a What population should he use? Give a reason for your answer.

  b Why might he not want to use a census of the house prices?

2 A town council wants to know what people think about the plan to build a new shopping centre.

  It decides to take an opinion poll of residents' views.

  a Give one reason why the council should not take a census.

  b From what population should it take a sample. Give a reason for your answer.

3 A market research company is going to conduct a national opinion poll.

  They want to find out what people think about current licensing laws (the times that pubs and clubs can open and close).

  Give **two** reasons why it would not be a good idea to carry out a census.

4 A market research company wants to find out customers' views about a new mobile phone shop that has just opened.

  Should they take a census or survey a sample of their customers?

  Give a reason for your answer.

EQ 5 The manager of a factory wants to carry out a survey to find out the workers' views on the menu in the factory canteen.

  Should the manager take a sample survey or a census?

  Give **two** reasons for your answer.

6 A city council is trying to decide where it should build a new school.

  Should they take a sample survey or a census?

  Give a reason for your answer.

Tabulation

5.1 Tally charts and frequency tables

Quick reminder

When data is collected it needs to be organised so that it is easy to read. A frequency table has three columns, one for listing the items which have been collected, one for the tally marks, and one to record the frequency of each item. Tally marks are grouped into fives.

Sport	Tally	Frequency
Football	ЖЖ ЖЖ ЖЖ ЖЖ ЖЖ	25
Rugby	ЖЖ ЖЖ ЖЖ III	18
Tennis	ЖЖ ЖЖ II	12
Basketball	ЖЖ ЖЖ ЖЖ ЖЖ II	22

rcise 5A

EQ 1 During activities week, the students at one school went on a trip to either a theme park, a cinema or an ice rink. Sahil asked 20 of his friends where they went. The results are listed below:

ice rink, theme park, cinema, ice rink, theme park, theme park, theme park, cinema, ice rink, ice rink, cinema, theme park, theme park, ice rink, ice rink, theme park, ice rink, theme park, theme park, cinema

a Copy and complete the frequency table.

Activity	Tally	Frequency
Theme park		
Cinema		
Ice rink		

b Which activity was the most popular?

EQ 2   Here are the colours of 30 cars in a car park.

red	silver	silver	black	silver	red
silver	black	red	black	white	black
black	white	red	silver	red	silver
black	blue	white	silver	black	silver
blue	silver	blue	white	red	black

a   Copy and complete the frequency table to show the colours of cars.

Colour of car	Tally	Frequency
Red		
Silver		
Black		
White		
Blue		

b   Which car colour was the most frequent?

c   Which car colour was the least frequent?

d   What percentage of cars were white?

EQ 3   In a survey students were asked to choose their favourite subject from a list of four: Mathematics, English, PE or Art. Here are the results:

Mathematics	PE	Mathematics	Art	PE	PE	Mathemat
English	Mathematics	English	PE	Mathematics	Mathematics	Mathemat
PE	Art	PE	Mathematics	English	PE	PE
Art	Mathematics	Art	PE	Art	Art	Mathemat

a   Copy and complete the frequency table to show favourite subjects.

Favourite subject	Tally	Frequency
Mathematics		
English		
PE		
Art		

b   Which subject was the most popular?

c   Which subject was the least popular?

d   How many students were in the class?

e   What fraction of students liked English?

EQ 4 A survey was carried out into the favourite drinks of some students.

Here are the results:

water	cola	milk	milk	lemonade
milk	water	water	cola	cola
lemonade	cola	cola	water	lemonade
cola	lemonade	milk	cola	cola
lemonade	lemonade	cola	lemonade	milk

a Copy and complete the frequency table to show favourite drinks.

Favourite drink	Tally	Frequency
Water		
Cola		
Lemonade		
Milk		

b What was the least popular drink?

c What fraction of the students like milk. Give your answer in its simplest form.

d What percentage of students liked cola?

EQ 5 Members of a class were asked how many pets they had. These are their answers:

1	2	1	2	2	3
6	5	1	5	2	2
1	4	1	4	2	2
2	4	1	3	1	3
3	3	2	3	2	4

a Draw a frequency table to show this information.

b What was the most common number of pets?

c What was the total number of pets?

d How many students had less than three pets?

EQ 6 Members of a class are asked how many brothers they have. These are their answers:

0	1	1	2	1
1	3	2	2	1
0	1	1	0	1
0	0	1	0	1

a Draw a frequency table to show this information.

b How many students had three brothers?

c How many students had less than two brothers?

7 Esme asked the other children in her class how many pets they had. The responses are listed below:

2, 4, 0, 0, 0, 3, 4, 2, 1, 3, 0, 1, 0, 2, 0, 2, 3, 1, 2, 2, 4, 3, 2, 0, 0, 1, 0

a Design and complete a tally chart to present this information. Include a frequency column.

b How many children had no pets?

c How many children did Esme ask?

5.2 Grouped frequency tables

Quick reminder

When the data to be collected has a wide range of values, with few values likely to be the same, the data is sorted into groups or classes. These are called **class intervals.**

Exercise 5B

1 Emily wanted to find out how much time her friends spent completing homework each week. She asked them to record the number of hours they spent completing homework over one week. The results are listed below:

1, 1, 2, 2, 3, 4, 4, 4, 5, 6, 6, 7, 7, 7, 7, 7, 8, 8, 8, 9, 9, 9, 9, 9, 11, 11, 12, 14, 15, 15, 16, 18, 19, 19

a Copy and complete the table.

Number of hours	Frequency
0–6	
7–12	
13–18	
19–24	

b How many people spent 13–18 hours completing homework?

c How many people spent 6 hours or less completing homework?

d How many people spent more than 13 hours completing homework?

2 The data shows the number of emails that 40 people in an office received during a working day.

23	23	26	29	22	25	44	46
31	31	17	25	34	49	45	21
39	42	35	41	40	23	44	29
46	27	11	17	12	21	34	30
42	49	13	18	14	32	37	32

a Copy and complete the frequency table below:

Number of emails	Tally	Frequency
11–15		
16–20		
21–25		
26–30		
31–35		
36–40		
41–45		
46–50		

b Tammy also decided to put the data into a grouped frequency table. Copy and complete Tammy's frequency table.

Number of emails	Tally	Frequency
1–20		
21–40		
41–60		

c Give a reason why the original frequency table might be more useful than Tammy's frequency table.

d Give a reason why Tammy's table might be more useful than the original frequency table.

EQ 3 Mr Jones gave some students a mental arithmetic test. There were 15 questions. The results are shown in the table below:

Marks in test	Frequency	Marks in test	Frequency
0	1	8	6
1	0	9	4
2	1	10	7
3	0	11	6
4	2	12	10
5	3	13	8
6	3	14	7
7	4	15	4

a How many students took the mental arithmetic test?

b What percentage of students scored 10 or more marks in the mental arithmetic test?

The data is then put into a grouped frequency table.

Number of marks in mental arithmetic test	Frequency
0–3	
4–7	
8–11	
12–15	

c Copy and complete the frequency table.

d Give one advantage of using the original frequency table.

e Give one advantage of using the grouped frequency table.

EQ 4 William measured the height, to the nearest centimetre, of 20 people. His results are listed below:

140, 143, 147, 147, 150, 150, 151, 154, 155, 156, 158, 159, 161, 162, 165, 166, 167, 170, 174, 177

Height (h) in cm	Frequency
$140 \leqslant h < 150$	
$150 \leqslant h < 160$	
$160 \leqslant h < 170$	
$170 \leqslant h < 180$	

a Copy and complete the grouped frequency table.

b How many people are over 160 cm tall?

c How many people had a height of less than 150 cm?

d Which group would someone with a height of 170 cm go in?

EQ 5 Molly measured the masses, to the nearest kilogram, of some students. Her results are listed below:

47   47   48   50   51   51   54   57   58   58   59   60

61   64   64   64   67   67   69   70   72   73   74

a Copy and complete the grouped frequency table.

Masses (m) in kg	Frequency
$45 \leqslant m < 50$	
$50 \leqslant m < 55$	
$55 \leqslant m < 60$	
$60 \leqslant m < 65$	
$65 \leqslant m < 70$	
$70 \leqslant m < 75$	

b How many people had a mass of less than 60 kg?

c Which group would someone with a height of 55 kg go into?

6 Pedro collected data about the pocket money that the other students in his class received in one week. His results are listed below. All the amounts are in euros.

5.00   11.25   6.65   8.80   12.00   13.20   13.55   12.80

8.50   12.25   5.00   10.00   14.40   9.80   10.20   7.80

15.00   10.45   6.30   9.60   11.60   10.00   8.70   6.75

Design a grouped frequency table to illustrate this data.

Choose your class limits class so that you have four or five equal intervals.

7 Some books were weighed. This is the list of their weights to the nearest gram.

52	43	44	34
23	26	27	29
24	21	37	39
29	30	37	47
41	43	44	20

a Copy and complete the grouped frequency table.

Weight w (nearest gram)	Tally	Frequency
$20 \leqslant w < 25$		
$25 \leqslant w < 30$		
$30 \leqslant w < 35$		
$35 \leqslant w < 40$		
$40 \leqslant w < 45$		
$45 \leqslant w < 50$		
$50 \leqslant w < 55$		

b How many books weighed less than 30 grams?

c In which class interval are the most books?

d Give one disadvantage of grouping data in this way.

To simplify the data the 20 values are regrouped as shown.

Weight w (nearest gram)	Tally	Frequency
$20 \leqslant w < 30$		
$30 \leqslant w < 40$		
$40 \leqslant w < 50$		

e Give two reasons why the first table is a better form of grouping than the second.

8 The table shows the time to the nearest minute for some students to complete a piece of homework.

Time, t minutes	Frequency
0–19	3
20–24	5
25–29	7
30–34	10
35–49	4
50–	2

a What are the class limits of the second class interval?

i Lower limit

ii Upper limit

b Why has the table got a class interval 50– ?

c Write down one reason why the class intervals are of varying width.

d Mrs Roberts believes that the students should be able to complete the homework well in under 35 minutes. Does the table support her belief? Give a reason for your answer.

5.3 Two-way tables

Quick reminder

A two-way table lets you show two variables at the same time.

Exercise 5C

EQ 1 The two-way table shows the number of slices of toast and the number of sausages that 40 students bought at breakfast one morning in the school canteen.

		Number of slices of toast			
		0	1	2	3
Number of sausages	0	5	7	2	1
	1	1	12	4	2
	2	2	2	1	1

a How many students bought 2 sausages and 3 slices of toast?

b How many students bought exactly one sausage?

c How many students bought at least one slice of toast?

d Look at the '5' in the table. How does the breakfast of these 5 students differ from the other 35 students in the table?

EQ 2 The two-way table shows the hair colour and gender of students in Mr Khan's tutor group.

		Hair colour				
		Brown	Blonde	Black	Ginger/auburn	Total
Gender	Boys	6	7	3	0	16
	Girls	7	4	1	2	14
	Total	13	11	4	2	30

a How many girls have black hair?

b How many boys have blonde hair?

c How many girls have neither brown nor black hair?

d What is the least frequent hair colour?

e How many students are in Mr Khan's tutor group?

f What percentage of student's in Mr Khan's tutor group have ginger/auburn hair?

EQ 3 The two-way table shows the number of televisions per house of 30 houses on a street and the number of cars per house. No household has more than 3 televisions or 3 cars.

	0 televisions	1 televisions	2 televisions	3 televisions
0 cars	1	1	2	0
1 car	0	3	5	1
2 cars	0	4	5	2
3 cars	0	1	3	2

a How many houses have exactly three cars and two televisions?

b How many houses have three cars?

c How many houses have two televisions?

d What percentage of households have exactly one television and one car?

EQ 4 100 students each study one of three option subjects.

The two-way table shows some information about these students.

	RS	German	History	Total
Female	35			53
Male		17		47
Total	47	22		100

a Copy and complete the two-way table.

b How many male students study History?

c How many female students study German?

d How many students study History?

e What percentage of students are female and study RS?

EQ 5   The two-way table gives some information about the lunch arrangements of 60 students.

	School lunch	Packed lunch	Other	Total
Female			6	32
Male	21	7		
Total	38			60

a   Copy and complete the two-way table.

b   How many female students have a packed lunch?

c   How many male students do not have a school lunch or packed lunch?

d   How many students have a packed lunch?

e   What fraction of students have something other than a school or packed lunch?

EQ 6   Alice is collecting some information as to whether the students in two classes watched football on television yesterday.

There are 61 students in total.

29 of the students are male.

24 females watched football.

8 boys did not watch football.

a   Use this information to copy and complete the two-way table.

	Male	Female	Total
Watched football			
Did not watch football			
Total			

b   How many female students did not watch football?

c   How many students did not watch football?

d   What percentage of male students did watch football?

EQ 7   A large LEA compiled the table below which illustrates the percentage of children who got their first, second or third choice secondary school in different parts of the authority. Some children did not get any of their choices.

Region	First Choice (%)	Second Choice (%)	Third Choice (%)	None of their choices (%)
North	83	15	2	0
West	84	12	3	1
South East	76	11	12	2
South West	81	3	13	3
East	89	10	1	0

a   What percentage of children in the West region got their second choice of school?

b   Select the region where the students have the greatest chance of getting the school of their choice.

c   Give a reason why the South East region does not total 100%.

6 Diagrammatic Representation

Pictograms, line graphs and bar charts

Quick reminder

A **pictogram** is a frequency table in which frequency is represented by a repeated symbol. The symbol itself usually represents a number of items, but can also represent just a single unit. The key tells you how many items are represented by a symbol.

A **bar chart** consists of a series of bars or blocks of the same width, drawn either vertically or horizontally from an axis. The heights or lengths of the bars always represent frequencies.

Line graphs are usually used to show how data changes over time. One such use is to indicate trends, for example, whether the Earth's temperature is increasing as the concentration of carbon dioxide builds up in the atmosphere, or whether a firm's profit margin is falling year-on-year.

Exercise 6A

EQ 1 The pictogram shows the numbers of days, in February, with more than one hour of sunshine for three cities: Southampton, Glasgow and Belfast.

The information for Swansea is missing from the pictogram.

Days in February with more than 1 hour of sunshine

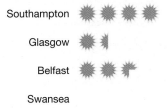

Southampton ✹ ✹ ✹ ✹

Glasgow ✹ ⟩

Belfast ✹ ✹ ✹

Swansea

Key ✹ represents 4 days

In the same month of February, Swansea had 8 days with more than 1 hour of sunshine.

a Copy and complete the pictogram.

b Write down the city that had the most days with more than 1 hour of sunshine in February.

c Write down the number of days with more than 1 hour of sunshine in Belfast.

Questions 2, 3, 4 and 5 relate to the data in the table below, which shows the number of cars caught speeding while driving past Frank's school last week.

Day	Mon	Tue	Wed	Thu	Fri	Sat	Sun
Frequency	70	59	41	20	40	52	21

2
  a Draw a pictogram to illustrate this data. Use a key of one car-shaped symbol to equal 20 cars.

  b Explain why a key of one symbol to equal 2 cars is not a good idea.

3
  a Draw a suitable bar chart to illustrate this data.

  b Explain why you think Monday had the most cars caught speeding and why the numbers on Tuesday, Wednesday and Thursday went down.

  c Give one advantage of drawing a bar chart rather than a pictogram for this data.

4
  a Draw a line graph to illustrate this data.

  b Can you use your line graph to predict the number of cars that will be caught speeding while driving past Frank's school next Monday? Give a reason for your answer.

5 Would the data about speeding cars, above, be best represented in a pictogram, a bar chart or a line graph?

Explain your answer.

Questions 6, 7, 8 and 9 relate to the data in the table below, which shows the number of beefburgers sold by a burger van at a week-long music festival.

Day	Sat	Sun	Mon	Tue	Wed	Thu	Fr
Frequency	900	1050	700	800	800	950	12(

6
  a Draw a pictogram to illustrate this data. Use a key of one circle to equal 100 beef burgers.

  b Explain why a key of one circle to equal 100 beefburgers is a good idea.

7 Explain why you think that most burgers were sold on the Friday.

8 Now draw a suitable bar chart to illustrate the same data.

9
  a Draw a line graph to illustrate this data.

  b Can you use your line graph to predict the number of beefburgers sold by the burger van on the next Saturday? Give a reason for your answer.

10 For her business plan, Michelle has produced a bar chart to compare the average number of flowers sold per week in her shop over the last two years.

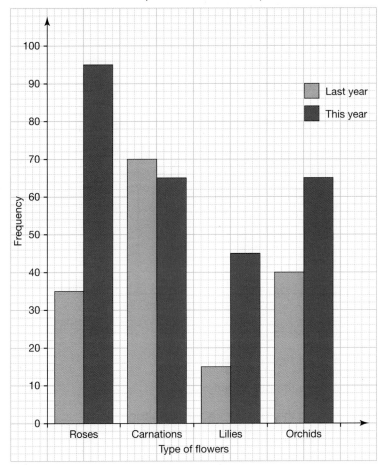

a Which flower was most popular last year?

b Which flower was most popular this year?

c How many roses were sold last year?

d How many more lilies were sold this year than last year?

E

EQ 11 The pictogram below shows the number of text messages received by four friends last week.

A03

Name		Number of texts received
Jason		
Amir		
Glynis		10
Aisha		8
	Total	

Key: = 5 text messages.

a How many texts did Amir receive?

b Copy and complete the pictogram.

c How many texts did the four friends receive altogether?

6.2 Pie charts

Quick reminder

Each category of data is represented by a sector of the pie chart. The angle of each sector is proportional to the frequency of the category it represents.

A pie chart cannot show individual frequencies, like a bar chart can. It can only show proportions.

Exercise 6B

F

1 Copy the basic pie chart shown here and complete it with the data given in the table below.

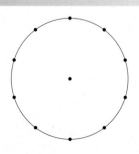

The favourite yoghurt flavours of 10 people.

Flavour	Vanilla	Strawberry	Toffee	Plain
Frequency	5	1	3	1

 2 The table shows the favourite ice cream flavours of students in the school canteen. **AO3**

Flavour	Vanilla	Strawberry	Toffee	Plain	Other
Frequency	32	6	17	3	2

Draw a pie chart to represent this information.

3 The pie chart below shows the proportions of the different number of sick days taken off by 48 staff in the same company last year. **AO3**

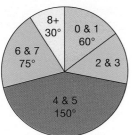

a What is the angle of the sector representing 2 and 3 sick days off?

b How many people took 4 or 5 sick days off?

4 A computer company's profits increase from £1 230 000 in 2010 to £1 345 000 in 2011. Sue is asked to draw two comparative pie charts to illustrate this increase. If she chooses a radius of 25 cm for the 2011 graph, calculate the radius she should use for the 2010 graph. **AO3**

EQ 5 The proportions of elm, beech and oak trees in Greg's Wood, in 1955 and 1995, are shown in the pie charts below.

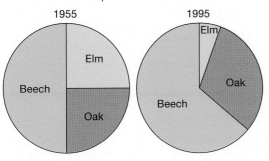

In the period from 1955 to 1995, most elm trees in Wales were killed by a fungus.

a Write down one feature shown in the pie charts that suggests that Greg's wood could be in Wales.

b Glynis says, 'These pie charts show that the number of beech trees in Greg's Wood has increased in the period from 1955 to 1995.

Glynis is **wrong**. Explain why.

6.3 Misleading graphs

Exercise 6C

1 The graph below shows the favourite types of TV programmes when 250 students were surveyed.

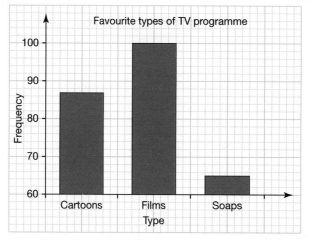

a Explain what is misleading about the graph.

b Redraw the graph to make it less misleading.

2 The graph below shows the changes in petrol prices over a four-year period.

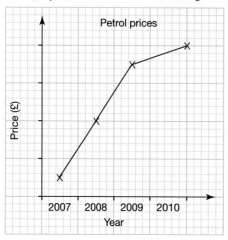

Explain what is misleading about the graph.

EQ 3 This graph appeared in a fishing magazine.

Number of fishing licences sold.

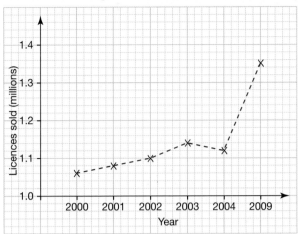

Give two reasons why the graph is misleading.

4 The table below shows the number of shoes sold in a shoe shop last week.

Day	Mon	Tue	Wed	Thu	Fri	Sat	Sun
Number of shoes sold	22	19	18	24	34	46	21

a Draw this information in a bar chart to make it a misleading bar chart. **AO3**

b Explain what your bar chart shows to someone who is not good at spotting a misleading bar chart. **AO4**

5 The table below shows the amount of petrol sold at a petrol station last week.

Day	Mon	Tue	Wed	Thu	Fri
Petrol sold (× 1000 litres)	400	340	200	180	550

a Draw this information in a line graph to make it look as though no petrol was sold on Thursday. **AO3**

b Explain how you made the line graph look misleading. **AO4**

EQ 6 Craig puts an advert in a local newspaper to encourage more customers to his shop.

He counts the number of customers the week before, the week during and the week after the advert is in the paper.

He draws this accurate 3D pie chart to show the results.

Number of customers

Craig says 'look at the pie chart! It is obvious that the advert was a success.'

Explain why this pie chart is misleading and shows that Craig is **wrong**.

6.4 Choropleth maps

Quick reminder

Choropleth maps are maps in which areas are shaded differently, to illustrate a distribution.

Every map should have a 'key' that makes sense of the data. Remember to study this carefully before answering a question.

Exercise 6D

1 The Choropleth map shows the car thefts per 1000 people per country per year in Europe.

AO3

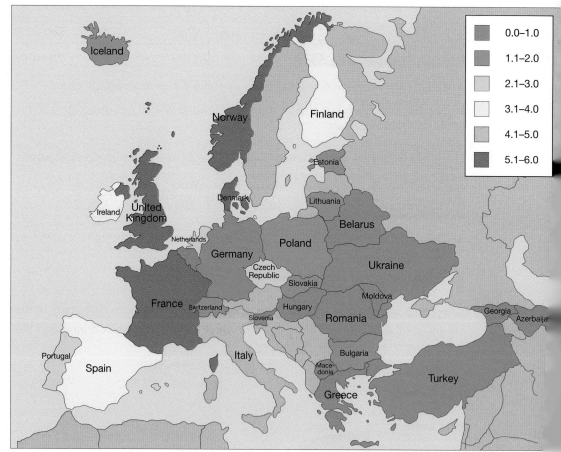

a Which is the only country to have between 4.1 and 5.0 cars stolen per 1000 people?

b How many cars per 1000 people are stolen in Spain?

c How could you improve on the way the information is displayed on this map?

2 Part of a coral reef is subdivided into square sections.

The number of different species of fish passing each square in an hour is shown in the table below.

AO4

24	20	23	22	18	19
26	23	19	21	10	18
39	33	30	22	9	6
25	21	24	21	17	19

a Use the key to produce a Choropleth map illustrating the data.

AO3

Key

b Part of the coral reef has been damaged by fishermen using dynamite.

Draw a line around the area where you think the coral has been damaged.

c Explain your answer to part **b**.

d One area is regularly visited by divers who feed the fish they see.

Draw a line around this area.

e Explain your answer to part **d**.

3 A forest is subdivided into square sections.

10	6	3	3	3	9
13	12	4	4	10	11
11	19	14	12	14	13
15	19	12	12	12	10

The number of different species of birds in each section is shown in each square of the table.

AO4

a Use the key to produce a Choropleth map illustrating the data.

AO3

Key

b Part of the forest has been cut down and the trees removed for their wood.

Draw a line around the area where you think the trees have been cut down and removed.

c Explain your answer to part **b**.

d One area of the forest has very fertile soil and a lot of trees grow there.

Draw a line around this area.

C

EQ **4** The Choropleth map shows the percentage change of population in Scotland from 2008 to 2009.

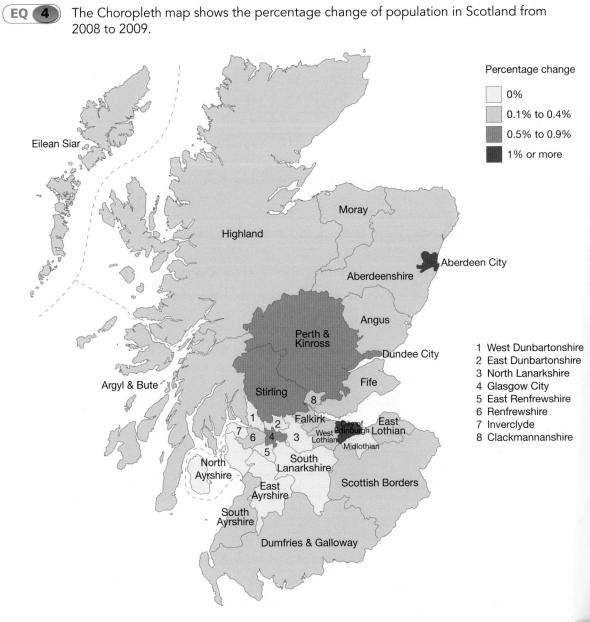

Percentage change

☐	0%
☐	0.1% to 0.4%
☐	0.5% to 0.9%
■	1% or more

Eilean Siar

Moray

Highland

Aberdeen City

Aberdeenshire

Angus

Perth & Kinross

Dundee City

Argyl & Bute

Fife

Stirling

8

1 West Dunbartonshire
2 East Dunbartonshire
3 North Lanarkshire
4 Glasgow City
5 East Renfrewshire
6 Renfrewshire
7 Inverclyde
8 Clackmannanshire

Falkirk

1

7 6 4 2 3

City of Edinburgh East Lothian

West Lothian

5

Midlothian

North Ayrshire

South Lanarkshire

East Ayrshire

Scottish Borders

South Ayrshire

Dumfries & Galloway

a Which region has the highest population change? A

b Give a reason for your answer to part **a**. A

c What is the percentage population change of Inverclyde?

d How could you improve on the way the information is displayed on this map? A

5 Each dot on the diagram below represents a child playing in the school playground.

AO4

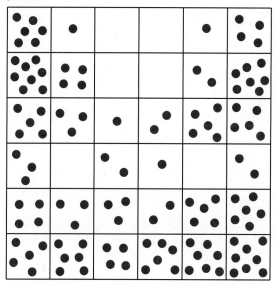

a Use the key to produce a Choropleth map illustrating the data.

AO3

Key

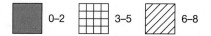

0–2   3–5   6–8

b Someone has thrown a stink bomb. Put a cross (X) where you think it landed.

c Explain your answer to part **b**.

d Draw an arrow on your Choropleth map to show which way you think the wind is blowing.

e Explain your answer to part **d**.

EQ 6 A forest is divided into 16 regions of equal size.

The owls' nests in each region are counted and the numbers are written in the squares, as shown in the following diagram.

0	1	0	0
2	3	6	2
1	5	8	1
0	1	1	2

Key | 2 | means there are two owls' nests in this region

a Use the information in the diagram to copy and complete this choropleth map.

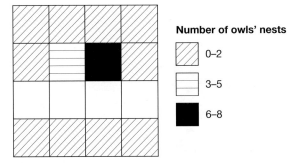

Number of owls' nests

- 0–2
- 3–5
- 6–8

b Describe how the owls' nests are spread across the forest.

6.5 Stem-and-leaf diagrams

Quick reminder

The numbers in a stem-and-leaf diagram may be left in the order they are given, but it is usual to rearrange them, to give an **ordered** stem-and-leaf diagram. The diagrams can be used to find measures such as the range and the median.

Exercise 6E

1 21 students try to guess when 10 seconds is up by pressing a computer key.

Their actual times are shown below:

10.8	11.4	9.7	11.5	10.5	11.3	11.1
12.3	10.9	11.5	10.6	10.8	12.0	11.3
11.6	11.5	12.1	10.6	10.9	11.1	11.9

a Construct a stem-and-leaf diagram to illustrate the data.

b Construct an ordered stem-and-leaf diagram to illustrate the data.

c Find the range.

d Write down the median value.

2 The masses of 31 tomatoes, in grams, are shown below:

Stem		Leaf
2		9 7 5 6
3		5 4 6 6 7 0 1 4 5 6 3
4		0 1 4 1 7 6 2 7 8
5		3 8 0 2 9
6		1 2

a Construct an ordered stem-and-leaf diagram.

b Write down a key.

c How heavy is the lightest tomato?

d What does the heaviest tomato weigh?

e What is the range of this data?

f Use your stem-and-leaf diagram to work out the median.

3 The back-to-back stem-and-leaf diagram below shows the times taken in seconds
to complete a puzzle successfully by two groups of different ages. **AO3**

Adults			14–16 year olds
8	2	0 1	
8 4	3	0 4 9	
9 4 2	4	5 6 7 7 7 8 8	
7 4 0 0	5	3 4 5 7 8	
8 8 7 5 3 1	6	0 2 4	
7 6 6 5 4	7	3 7	
7 4 3 2	8	2 4	
4 4 2	9	2	

Key: 3 | 4 means 34 seconds

a What was the shortest time taken by the 14–16 year olds?

b What was the longest time taken by the adults?

c Write down the median values of both data sets.

d Comment on how well each group performed. **AO4**

EQ 4 A nurse took the pulse rate of 23 patients.

The pulse rates are shown in the stem-and-leaf-diagram.

Pulse rates

6	7 8
7	2 3 4 4
7	5 5 5 5 8 9
8	0 2 3 3 4
8	8 8 9
9	0 4 4

Key: 6 | 7 means 67 beats per minute

a Write down the median pulse rate.

b Write down the mode pulse rates.

The sum of all the pulse rates was 1840.

c Work out the mean pulse rate.

6.6 Histograms and frequency polygons

Quick reminder

Histograms look similar to bar graphs but there are no gaps between the bars. They are represented in two formats: equal class-width intervals and unequal class-width intervals.

Frequency polygons are often used instead of histograms to compare two sets of data. Each point is plotted in the centre of the group and the points are joined up using a straight line.

Exercise 6F

1 The table below shows a grouped frequency distribution of the marks that a class of students achieved in their Spanish exam.

AO3

Spanish result (%)	40–45	46–50	51–55	56–60	61–65
Number of students	2	5	9	8	4

a Draw a frequency diagram to illustrate the results.

b Draw a frequency polygon on the same diagram.

2 A magazine carried out a survey of the ages of its readers.

The table shows the results of the survey.

Age, y (years)	Frequency
$5 \leqslant y < 15$	8
$15 \leqslant y < 25$	39
$25 \leqslant y < 35$	30
$35 \leqslant y < 45$	15
$45 \leqslant y < 55$	8

a Draw a frequency polygon for this data.

A

b The magazine editor is thinking of including an article about buying antiques. Do you think this is a good idea? Explain your answer.

A

3 The frequency polygon shows the ages of some people who attended a charity fun run.

A

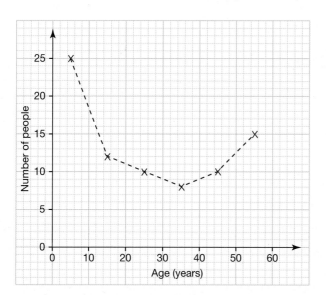

Hamish said, 'most of the fun runners were 5 years old'.

Explain why this might be wrong.

4 The mass, in kg, of the suitcases on a plane are shown below:

AO3

12.7	10.4	16.0	16.0	12.8	15.2
19.1	14.0	14.8	16.0	16.7	15.1
10.1	15.3	11.2	12.6	18.3	11.6
17.6	13.3	10.3	19.3	11.6	15.3
19.2	14.5	17.0	15.1	15.1	18.4
16.0	16.3	12.2	16.1	11.7	16.7
10.2	15.3	15.2	13.3	15.5	15.8

a Copy and complete the frequency table.

Mass, m	Tally	Frequency
$10 \leqslant m < 12$		
$12 \leqslant m < 14$		
$14 \leqslant m < 16$		
$16 \leqslant m < 18$		
$18 \leqslant m < 20$		

b Draw a histogram to illustrate the data.

c Write down the modal class.

d In which group does the median lie?

Hints and tips

Remember to use frequency density when drawing a histogram.

EQ 5 The number of merits gained by Year 9 students are summarised in the table below:

Number of merits, m	Frequency
$0 \leqslant m < 20$	24
$20 \leqslant m < 40$	52
$40 \leqslant m < 60$	40
$60 \leqslant m < 90$	33
$90 \leqslant m < 150$	15

a Draw a histogram for this data.

b Any student with more than 80 merits gets to go on a school trip.

Calculate an estimate of the number of students who get to go on a school trip.

EQ 6 The distances travelled by 100 different cars, each using 1 litre of petrol, are shown in the following histogram and table.

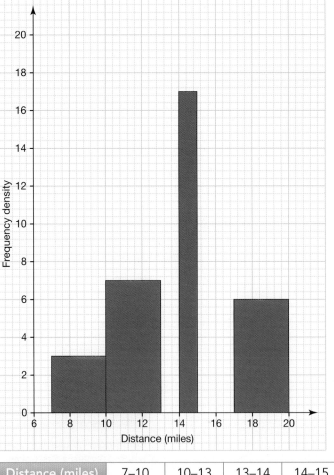

Distance (miles)	7–10	10–13	13–14	14–15	15–17	17–20
Frequency	9		18		17	

a Copy and complete the histogram and the table.

b Estimate the percentage of cars that travel between 12 and 16 miles on 1 litre of petrol.

EQ 7 The distances travelled by 200 vans, each using 1 litre of diesel fuel, are shown in the following histogram and table.

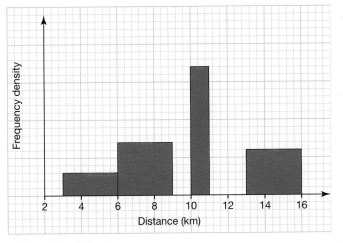

Distance (km)	3–6	6–9	9–10	10–11	11–13	13–16
Frequency	18		36		34	
Frequency density						

a Copy and complete the table.

b Copy the histogram. Work out the scale on the vertical axis.

c Complete the histogram by drawing the two missing bars.

d Estimate the percentage of vans that travel between 8 and 16 km on 1 litre of diesel fuel.

<table>
<tr><td>**6.7**</td><td># Cumulative frequency graphs</td></tr>
</table>

Quick reminder

A cumulative frequency graph can be used to find the interquartile range and the median.

Exercise 6G

1 160 students took a test. The cumulative frequency graph shows information about their marks.

AO3

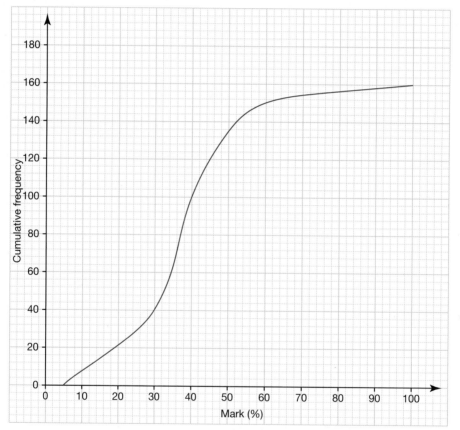

Work out an estimate for the interquartile range of their marks.

2 At a fundraising event, a game consists of spinning two large spinners and adding the scores. The results of the first 140 people to play the game are recorded in the table below:

AO3

Spinner score, s	Frequency
$1 \leqslant s < 20$	11
$21 \leqslant s < 40$	21
$41 \leqslant s < 60$	55
$61 \leqslant s < 80$	27
$81 \leqslant s < 100$	18
$101 \leqslant s < 120$	8

a Draw a cumulative frequency diagram to show the data.

b Use your diagram to estimate the median score.

c Use your diagram to estimate the interquartile score.

d Those who score more than 90 win a prize. What fraction of the people get a prize?

e Those who score less than 30 are given another turn. Approximately what percentage of the people are allowed to spin again?

3 Jordan records the number of goals scored in his school's football league using the table below:

AO3

Number of goals	0	1	2	3	4	5	6
Frequency	7	10	24	25	10	6	2

a Draw a cumulative frequency step polygon to illustrate the data.

b Find the median of the data.

c Find the Interquartile range of the data.

EQ 4 The length of time, in seconds, of 40 burps in a burping competition were recorded. A cumulative frequency diagram of this data is shown on the grid below.

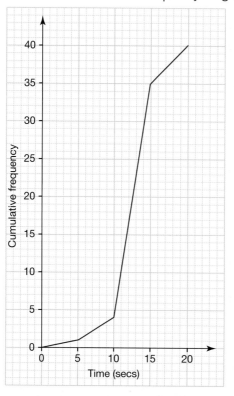

Use the diagram to find the limits between which the middle 50% of the times lie.

 EQ 5 For 100 days Gemma kept a record of how late her bus was.

The information is shown in the following table.

Time late (t mins)	Frequency
$0 \leqslant t \leqslant 2$	23
$2 < t \leqslant 4$	35
$4 < t \leqslant 6$	24
$6 < t \leqslant 8$	12
$8 < t \leqslant 10$	5
$10 < t \leqslant 12$	1

a Copy and complete the cumulative frequency table.

Time late (t mins)	Cumulative frequency
$0 \leqslant t \leqslant 2$	
$2 < t \leqslant 4$	
$4 < t \leqslant 6$	
$6 < t \leqslant 8$	
$8 < t \leqslant 10$	
$10 < t \leqslant 12$	

b Using the information in your table, draw a cumulative frequency diagram.

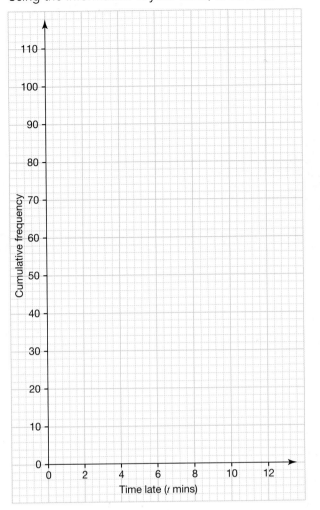

c Use your cumulative frequency diagram to find an estimate for the median time that the bus was late.

7 Measures of location

7.1 The mode

Quick reminder

The **mode** of a list of data is the number that occurs most often. So, for a frequency distribution, the mode is the number with the highest frequency.

Exercise 7A

1 Find the mode of each of these lists of data:

a 2 3 4 4 5 8

b 2 3 3 4 5 5 5 7

c 1 6 2 6 7 8 2 1 5 6 9

AO

2 A teacher asked 12 children how many brothers and sisters they had. The results were:

0 3 0 1 1 2 0 4 2 1 0 3

What was the mode?

AC

3 The cost of a carton of milk in ten supermarkets was:

85p 84p 90p 86p 84p 95p 86p 85p 87p 86p

What is the modal cost?

A

4 There are three different letters in the word BANANA.

Which is the modal letter?

A

5 Bernie does a survey on the makes of cars in the school car park. Here are her results.

Ford Ford Seat Mitsubishi Seat Seat Volvo Seat Vauxhall VW VW Vauxhall Vauxhall Vauxhall Seat Vauxhall Mitsubishi Seat Ford Mitsubishi Jaguar VW Ford VW VW Seat VW Volvo

What is the modal make of car?

A

6 For each of the frequency tables below, calculate the mode: **AO3**

a

x	3	4	5	6
f	14	10	7	3

b

x	4	6	8	10	12
f	2	11	12	14	6

c

x	5	10	15	20	25	30
f	6	12	11	15	12	5

7 The table below shows the number of goals scored last season by a football team: **AO3**

Number of goals	0	1	2	3	4	5	6	7
Frequency	8	15	12	5	2	0	0	1

What was the modal number of goals scored?

8 This bar chart shows the number of bedrooms in the houses in a street. **AO3**

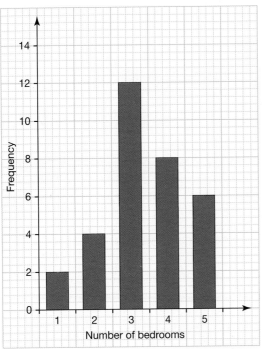

There are four two-bedroom houses.

What is the modal number of bedrooms?

9 Three numbers add up to 10. **AO3**

If the mode is 4, what are the three numbers?

10 Six numbers add up to 30. Two of the numbers are both 5. **AO3**

If the mode of the six numbers is 4, what are the six numbers?

EQ 11 The stem-and-leaf diagram shows the mass of young mice in a pet shop.

```
4 | 6  6  8
5 | 0  0  2  5  5  6  8  9
6 | 1  2  3  3  4  8  8  8  9
7 | 0  2  2  5  5  5  5  7  9
8 | 0  3  4
```

Key: 5|0 means 50 g

Write down the modal mass.

EQ 12 The following table shows the height of seedlings in a greenhouse.

Height, h (cm)	Frequency
$10 \leqslant h < 12$	16
$12 \leqslant h < 14$	41
$14 \leqslant h < 16$	36
$16 \leqslant h < 18$	34
$18 \leqslant h < 20$	12

Write down the modal class.

7.2 The median

Quick reminder

The median of a list of data arranged in ascending or descending order is the middle number.

When there is an even number of data items find the middle pair, add them together and divide by two.

If there are n items in a list, the position of the median can be found by working out $\dfrac{n + 1}{2}$.

i.e. the median is the $\dfrac{n + 1}{2}$th item in the list.

Exercise 7B

1 Find the median of each of these lists of data:

a 2 3 4 5 6 8 10

b 2 3 3 3 4 6 7 7 8

c 1 10 2 13 7 8 2 1 5 11 9

d 5 7 9 10 14 15 16 20

e 4.1 3.6 2.2 5.1 8.4 3.7

2 The prices of nine books that Ari is looking at are:

£8.50   £7.99   £5.40   £6.35   £7.99   £6.99   £3.50   £5.99   £6.50

a What is the modal price?

b What is the median price?

AO3

3 A social worker did a survey of 12 families. The number of people in the families was:

2   5   1   6   4   3   4   2   1   4   3   4

What is :

a The modal number?

b The median?

AO3

4 The number of children in 10 classes is:

22   25   26   25   23   21   20   21   22   25

What is :

a The modal number?

b The median?

AO3

5 For each of the frequency tables below, calculate the median:

AO3

a

x	1	2	3	4
f	1	5	6	3

b

x	14	16	18	20	22
f	2	7	8	14	7

c

x	15	20	25	30	35	40
f	16	8	3	2	2	1

6 The table below shows the number of goals conceded by a hockey team last season:

AO3

Number of goals	0	1	2	3	4	5
Frequency	11	8	3	1	0	1

What is:

a The modal number of goals?

b The median number of goals?

7 Ben works as a waiter. One Saturday night, he records the total bill for each table he serves and puts the information in this stem-and-leaf table:

AO3

Stem		Leaf
2		8 9
3		1 3 3 5 7 8
4		2 2 5 6 9 9 9
5		3 4 4 6
6		2 8
7		1 2 2

Key: 2 | 8 represents £28

a How many tables did he serve?

b What was the modal bill?

c Find the median bill.

8 Five numbers have a mode of 4 and a median of 6.

AC

Write down any of the five numbers you can and explain what you can tell about the value of the others.

9 Six friends compare their pocket money.

A

The median is £6.50.

What can you tell about the third and fourth largest amounts of pocket money?

EQ 10 A survey was done to count the number of eggs in sparrows' nests in two different regions. Each region had the same area.

The data have been recorded in this table.

Number of eggs	Frequency: region A	Frequency: region B
4	7	3
5	14	7
6	15	18
7	9	23
8	4	20
9	1	6
Total		

a Write down the median number of eggs counted in a nest in region A.

b Write down the median number of eggs counted in a nest in region B.

c Which region, A or B, do you think is nearest to easily available food? Explain your answer.

EQ 11 The table gives information about the neck sizes, in inches, of 20 shirts.

Neck size (inches)	Frequency
14	2
$14\frac{1}{2}$	7
15	6
$15\frac{1}{2}$	5

a Write down the modal neck size.

b Work out the median neck size.

c Work out the mean neck size.

d Which average would **best** describe the neck sizes of the 20 shirts?
Give a reason for your answer.

7.3 The mean

Quick reminder

$$\text{Mean} = \frac{\text{Sum of all the values}}{\text{Total number of values}}$$

For a frequency distribution, this becomes $\quad \text{Mean} = \dfrac{\text{Sum of } fx}{\text{Sum of } x}$

ercise 7C

1 Find the mean of each of these lists of data: AO3

a 1   2   4   4   6   7   11

b 3   4   4   4   5   6   6   7   15

c 10   2   13   7   8   2   1   5   11   9

d 4   7   9   10   14   15   16   21

2 Six girls use a social networking website MyFace. They each count how many
friends they have on the site and the results are: AO3

31   21   20   35   17   20

Find:

a the mean

b the mode

c the median

3 This table shows the number of boys and girls in each year in a school.

Year	7	8	9	10	11
Number of Boys	51	43	52	61	58
Number of Girls	46	57	64	59	44

Find:

a the mean number of boys in each year.

b the mean number of girls in each year.

c the mean number of pupils in each year.

4 For each of these sets of data, find:

i the mean   **ii** the mode   **iii** the median

a 3   3   3   4   5   5   12

b 9   3   9   7   5   9

c 7   1   3   4   2   1   5   2   3   2

5 Four boys are collecting conkers. The mean number of conkers they have is 7.

a What is the total number of conkers the four boys have?

b A fifth boy who only has two conkers joins them. What is the total number of conkers they have between the five of them?

c What is the mean number of conkers for the five boys?

6 Ariba goes on holiday for 7 days. The mean maximum temperature for the first 6 days is 89°F.

The maximum temperature on the last day is 103°F.

What is the mean maximum temperature for the whole week?

7 For each of the frequency tables below, calculate the mean:

a

x	4	5	6	7
f	1	3	4	2

b

x	20	40	60	80	100
f	2	5	8	3	2

c

x	20	25	30	35	40	45
f	1	3	3	1	0	2

8 Joon does a survey to see how many text messages people send. He asks 100 students how many texts they sent yesterday evening.

AO3

The results are shown in this table.

Number of texts	Frequency
0	6
1	25
2	31
3	24
4	14

Calculate the mean.

9 A town planner counts the number of people in 80 cars in a traffic survey.

AO3

The results are shown in this table.

Number of people	Frequency
1	3
2	21
3	18
4	20
5	12
6	2
7	3
8	1

Calculate:

a the mean

b the mode

c the median

EQ **10** A survey was done to count the number of eggs in sparrows' nests.

The data has been recorded in the following table.

Number of eggs (x)	Frequency (f)	fx
4	7	28
5	14	70
6	15	90
7	9	
8	4	
9	1	
Total		

Work out the mean number of eggs.

Give your answer correct to 1 decimal place.

7.4 Which average to use

Quick reminder

In certain circumstances, either the mean or the median or the mode can be a better 'average' to use than the other two. This may be because the person calculating the average wants to use it to prove something.

The mean can be affected badly by a few very large or very small items.

There can be more than one mode or none.

The average should be representative of all the data.

In this exercise, give your answers correct to 3 significant figures where necessary.

Exercise 7D

1 For each set of data, find:

i the mean   **ii** the mode   **iii** the median

a 2 2 9 10 11 12

b 2 3 4 6 7 19 19

c 1 1 15 16 18 19 20 20

For each set, state which average you feel represents the data best and give reasons.

2 If you collected data for each of these variables, what do you feel might be the best average to represent that data? Give reasons for your answers.

a Favourite flavour of crisps.

b Hours spent watching television yesterday.

c Time spent by patients waiting at a doctor's surgery.

3 A small firm employs ten people. Their annual salaries are:

£8000   £8000   £22 000   £23 000   £24 000

£25 500   £26 000   £27 500   £28 000   £90 000

What is:

a The mode?

b The median?

c The mean?

Which average would you say is most representative of this data and why are the other two not as useful?

D

4 A builder does a survey of 12 families. The numbers of people in the families are: **AO4**

2 5 1 6 4 3 4 2 1 4 3 4

What is:

a The mode?

b The median?

c The mean?

He is going to use the information to decide what size houses he should build. Which average – the mean, the mode or the median – do you think will be most useful to him and why?

5 I am a teacher talking to my Year 10 class of 13 pupils. They are all 15 years old, but I am 50 years old.   **AO4**

For our ages, what is:

a The mode?

b The median?

c The mean?

Which one of these averages would you say does not represent the data very well?

6 Fred buys a lottery ticket every week.   **AO4**

Over the last eight weeks, he has won £100 once, £10 three times and nothing on the other four weeks.

For these amounts, calculate:

a The mode

b The median

c The mean

Fred is trying to decide whether he should stop buying tickets. Which average do you feel is most useful to him?

7 In Diana's class, 6 girls have brown eyes, 4 girls have blue eyes, 2 girls have green eyes and one has grey eyes.   **AO4**

Which is the only average you can find for the colour of these girls' eyes?

8 Donald uses a spell checker to correct the mistakes he has made in his English coursework.   **AO4**

The number of errors on each page is:

2 5 12 13 8 45 11 9 10

For these numbers of errors, calculate if possible:

a The mode

b The median

c The mean

Giving reasons, say which of these averages best describes the data.

9 A chocolate manufacturer makes tubes containing small chocolate sweets.

AO4

It claims an average of 40 sweets in each tube.

Jerry decides to check this claim and counts the number of chocolates in 20 tubes with the following results:

Number of chocolates in a tube	Frequency
0	1
35	2
36	2
37	4
38	3
39	3
40	5

a For the number of chocolate sweets in a tube, calculate:

i the mean

ii the median

iii the mode

b Jerry then discovers that the empty tube had not originally been empty, but that the chocolates had been eaten by his little brother before Jerry could count them. He decides to discard this tube from his investigation and only use the data for the other 19 tubes.

i How will this affect the three averages you have found?

ii Which average is the manufacturer using and why?

EQ 10 Mr Truman's PE class and Mr Thom's PE class had a 'throw the cricket ball' competition.

The results are shown below.

Mr Truman's PE class: 31, 29, 47, 50, 27, 7, 33, 21, 37, 29, 47, 36, 22

Mr Thom's PE class: 36, 30, 33, 30, 37, 30, 35, 33, 31, 34

a Which 'average' best describes Mr Truman's PE class?

b Which 'average' best describes Mr Thom's PE class?

c Which class do you think is the best at throwing a cricket ball? Give a reason for your answer.

7.5 Grouped data

Quick reminder

When you use Mean = $\dfrac{\text{Sum of } fx}{\text{Sum of } x}$ for a grouped frequency table, you must use the mid-points of each class as your values of x. This gives an estimate of the mean.

If all the classes have the same width, the modal class is the class with the highest frequency.

Exercise 7E

EQ 1 The following table gives information about the length of time a random selection of 60 people took to text the sentence:

'The quick brown fox jumped over the lazy dog.'

Time taken, x, (seconds)	Number of people	Mid-point value	
$20 \leqslant x < 30$	18	25	
$30 \leqslant x < 40$	12		
$40 \leqslant x < 50$	6		
$50 \leqslant x < 60$	24		
		Total	

a Copy and complete the table.

b Work out the mean time taken.

2 Calculate an estimate of the mean of each of these grouped frequency tables, where x is a continuous variable.

AO3

Give your answers correct to 3 significant figures where necessary.

a

Class	Frequency
$1 \leqslant x < 5$	1
$5 \leqslant x < 9$	3
$9 \leqslant x < 13$	7
$13 \leqslant x < 17$	2
$17 \leqslant x \leqslant 21$	2

D

C

b

Class	Frequency
$10 \leqslant x < 20$	18
$20 \leqslant x < 30$	13
$30 \leqslant x < 40$	7
$40 \leqslant x < 50$	4
$50 \leqslant x \leqslant 60$	1

c

Class	Frequency
$1 \leqslant x < 5$	1
$5 \leqslant x < 10$	17
$10 \leqslant x < 15$	20
$15 \leqslant x < 20$	10
$20 \leqslant x \leqslant 25$	20

 3 The amounts spent by 30 customers at a supermarket were recorded and the following table was produced:

AO:

Amount spent, £x	Frequency
$0 < x \leqslant 20$	12
$20 < x \leqslant 40$	7
$40 < x \leqslant 60$	6
$60 < x \leqslant 80$	5

a Write down the modal class.

b Calculate an estimate of the mean.

 4 The amounts spent by 40 customers at a supermarket were recorded and the following table was produced:

A

Amount owing, £x	Frequency
$0 < x \leqslant 50$	2
$50 < x \leqslant 100$	13
$100 < x \leqslant 150$	15
$150 < x \leqslant 200$	6
$200 < x \leqslant 250$	3
$250 < x < 300$	1

Calculate an estimate of the mean.

5 50 students were asked how long they spent using a computer in one day.   **AO3**

The results were collected and presented in this table:

Length of time, t (hours)	Frequency
$0 < t \leq 2$	13
$2 < t \leq 4$	18
$4 < t \leq 6$	11
$6 < t \leq 8$	5
$8 < t \leq 10$	3

a Write down the modal class.

b Calculate an estimate of the mean.

6 100 students guess how long the headteacher's speech will last on Prize giving Day.   **AO3**

The guesses were grouped into the following table:

Length of time, t (mins)	Frequency
$0 < t \leq 4$	21
$4 < t \leq 8$	25
$8 < t \leq 12$	33
$12 < t \leq 16$	11
$16 < t \leq 20$	5
$20 < t \leq 24$	4
$24 < t \leq 28$	1

a Write down the modal class.

b Calculate an estimate of the mean.

7 Jess is investigating the length of books. She goes online and finds the number of pages in each of the week's top-selling books.   **AO3**

To present the data clearly, she puts the lengths into groups and produces the following table:

Number of pages, p	Frequency
$0 \leq p \leq 99$	12
$100 \leq p \leq 199$	37
$200 \leq p \leq 299$	33
$300 \leq p \leq 399$	18

Use Jess's table to estimate the mean.

8 Miss Spell records the marks that Year 9 get in their English exam and puts them in a table:

AO3

Marks, m	Frequency
11–20	1
21–30	0
31–40	26
41–50	29
51–60	34
61–70	12
71–80	2
81–90	1
91–100	1

Use the table to estimate the mean.

9 A doctor times how long 20 patients take to do a test on two consecutive days:

AO

Time, t (secs)	$10 < t \leqslant 15$	$15 < t \leqslant 20$	$20 < t \leqslant 25$	$25 < t \leqslant 30$	$30 < t \leqslant 35$
Frequency: Day 1	1	4	5	7	3
Frequency: Day 2	5	5	2	7	1

He uses the table to estimate the mean time on each day.

On which day was the estimate of the mean time lower, and by how much?

10 Jane estimates the mean for a grouped frequency table. She rubs out one of the frequencies and gives the table to her friend Jo. She tells Jo that the estimate of the mean was 8 and asks her to calculate the missing frequency.

A

Here is the table she gives to Jo:

t	$0 < t \leqslant 4$	$4 < t \leqslant 8$	$8 < t \leqslant 12$	$12 < t \leqslant 16$
Frequency	2		3	7

What is the missing frequency?

7.6 The geometric mean

The geometric mean of n numbers is the nth root of their product.

So the geometric mean of 4 and 9 is $\sqrt{4 \times 9} = \sqrt{36} = 6$

and the geometric mean of 1, 2, 8 and 16 is $\sqrt[4]{1 \times 2 \times 8 \times 16} = \sqrt[4]{256} = 4$.

Usually the geometric mean will not be a whole number, so you should write down the first five figures of your answer and then round it to 3 significant figures (or whatever the question says).

e.g. the geometric mean of 2, 4 and 9 is $\sqrt[3]{2 \times 4 \times 9} = \sqrt[3]{72} = 4.1601 = 4.16$ (to 3 sig. fig.)

Exercise 7F

1 Find the geometric mean of:　　　　　　　　　　　　　　　　　　**AO3**

   a 4 and 16

   b 1.4 and 2.9

   c 3, 8 and 9

   d 2.5, 4.7 and 6.4

   e 3, 6, 8 and 11

   f 2, 5, 13, 17 and 24

2 What would you multiply a sum of money (e.g. £60) by if you wanted to:　　**AO3**

   a increase it by 4%?

   b increase it by 3%?

   c increase it by 12%?

   d increase it by 2.5%?

   e decrease it by 5%?

   f decrease it by 13%?

3 Meghan invests some money for 5 years at 4% compound interest per year. By what percentage will the money have increased at the end of the five years?　**AO3**

4 On four consecutive days, the price of a share goes up by 4%, up by 2%, up by 5% and down by 3%. Use the geometric mean of the multipliers to find the average percentage change over the four days.　　**AO3**

5 A new DJ takes over a popular radio programme. During the next month the listening figures go up by 15%, and in the following two months they rise by 6% and 3%. Use the geometric mean of the multipliers to find the average percentage change over the three months.

AO3

6 Nancy is investigating inflation between 2004 and 2008. This table gives the percentage increase in the price of a certain chocolate bar each year for five years: AO3

Year	2004	2005	2006	2007	2008
Percentage rise	12%	9%	7%	3%	5%

Use the geometric mean to find the average percentage rise.

EQ 7 Warm-air Ltd produce electric fan heaters.

During the last four years the cost of the heating element has changed.

The following table shows these changes.

Year	Percentage price change
2007–2008	+14%
2008–2009	+5%
2009–2010	−3%
2010–2011	+7.5%

By what percentage has the heating element changed over this period?

8 Nancy finds this table, which shows how the price of a bottle of a certain drink has changed over time:

AO

Year	2004	2005	2006	2007	2008
Price	£1.10	£1.21	£1.28	£1.30	£1.35

The price in 2003 was exactly £1.

Work out the multipliers for each of the five years in the table and use the geometric mean to find the average percentage rise. Give your answer to one decimal place.

9 Using the geometric mean, the average percentage fall in the value of my car during the last two years was 28%.

If it fell 36% in the first year, by what percentage did it fall in the second year?

10 Michael writes down three numbers. He asks his friend Matthew to guess the numbers.

He tells Matthew that the mode of the numbers is 24 and the geometric mean is 12.

What are the three numbers?

8 Measures of spread

8.1 Box-and-whisker plots

Quick reminder

Here is an example of a box-and-whisker plot to show the ages of a sample of men:

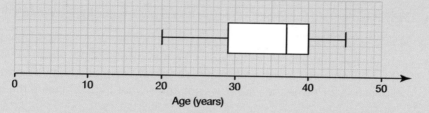

- The ends of the whiskers are 20 (the age of the youngest man) and 45 (the age of the oldest man).

- The lower quartile, Q1, is 29; this is where the box starts.

- The upper quartile, Q3, is 40; this is where the box ends.

- The median is 37 and this is shown by a line inside the box.

- The interquartile range is bigger if the data is more spread out. Here it is 40 – 29 = 11.

- The box plot shows that the sample is negatively skewed because there is more of the box on the left (negative) side of the median line.

Exercise 8A

1 Find the mode of each of these lists of data: AO3

	Smallest value	Lower quartile	Median	Upper quartile	Largest value
a	5	17	23	28	35
b	16	22	26	34	56
c	21	30	35	42	51

2 The box plot in the introduction shows the ages of a sample of 100 men. A similar sample of 100 women was taken and the results were: youngest age = 24, oldest age = 40, lower quartile = 27, median = 29, upper quartile = 34.

B

a Copy the diagram for the men and draw a box plot for the women using the same scale above it.

AO3

b Comment on the difference between the two distributions, making reference to the medians, the interquartile ranges and the skewness, and interpreting the comparisons for someone who does not know what these statistics mean.

AO4

3 In a town there are two rival garages – Green's Garage and White's Garage. This box plot shows the price of cars at Green's Garage.

The same data for prices of cars at White's Garage is: cheapest car £2500, most expensive £5000, lower quartile £3000, median £3300 and upper quartile £4000.

a Copy the diagram for Green's Garage and draw a box plot for White's Garage above it, using the same scale.

AO

b Comment on the differences between the two distributions, making reference to the medians, the interquartile ranges and the skewness, and interpreting the comparisons for someone who does not know what these statistics mean.

AO

4 The table below shows the hours of sunshine each day recorded over a year in two different resorts.

	Least hours	Lower quartile	Median	Upper quartile	Most hours
Resort A	5	7	9	11	13
Resort B	1	5	9	13	17

a Draw box plots to compare both sets of data.

b Comment on the differences between the distributions.

 5 Hannah says that she expects the data she is going to collect about the pocket money of Year 7 girls will be symmetrical. Do you think Hannah is correct? Explain your answer.

6 The table below shows the marks scored by 100 candidates in a maths exam.

Mark	No. of students
11–20	3
21–30	10
31–40	13
41–50	20
51–60	23
61–70	15
71–80	12
81–90	4

a Draw a cumulative frequency curve to show the data.

b Use your graph to estimate the median mark and the upper and lower quartiles.

c The lowest mark was 5 and the highest was 87. Draw a box plot to show the distribution of marks.

 Nick has photocopied two box plots showing the September midday temperatures of Torrevieja and Granada, which are two towns in Spain.

AO4

One town is next to the sea. The other is high in the mountains.

Unfortunately, the writing on the plots is not clear enough for Nick to read it.

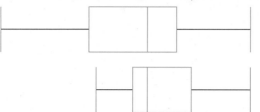

a Which box plot is the one for Granada? Explain your choice.

b Compare the midday temperatures shown by the two box plots.

EQ 8 27 track and field athletes each ran 200 metres. The time was recorded to the nearest second.

The stem-and-leaf diagram shows this information.

Athletes' times

2	2 5 5 6 6 6 8 8 9
3	1 2 3 3 3 4 6 6 7 7 7 8
4	2 2 5 5 5
5	5

Key: 3 | 1 = 31 seconds

a Use the information in the stem-and-leaf diagram to complete the following table.

Lowest value	
Lower quartile	28
Median	
Upper quartile	38
Highest value	

b Identify any outliers for the times of the athletes.

c On a copy of the grid, draw a box plot to show the distribution of the times of the athletes.

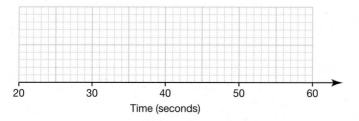

A

9 Here are four cumulative frequency curves and four box plots.

AO3

Match each cumulative frequency curve with a box plot.

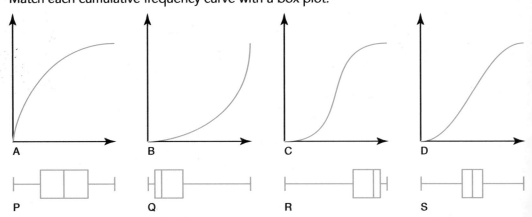

8.2 Variance and standard deviation

Quick reminder

These two quantities measure spread: the more variation there is within a set of data, the bigger they will be.

e.g. Set A consists of the six numbers   5, 5, 6, 6, 6, 7

Set B consists of the six numbers   2, 4, 5, 6, 7, 9

The standard deviation and variance for Set B will be bigger than for Set A because the numbers in Set B are more spread out.

The formulae you should use for standard deviation are:

For a list of numbers: Standard deviation $= \sqrt{\dfrac{\Sigma x^2}{n} - \left(\dfrac{\Sigma x}{n}\right)^2}$ or $\sqrt{\dfrac{\Sigma x^2}{n} - (\overline{x})^2}$

For a frequency distribution: Standard deviation $= \sqrt{\dfrac{\Sigma f x^2}{\Sigma f} - \left(\dfrac{\Sigma x}{\Sigma f}\right)^2}$ or $\sqrt{\dfrac{\Sigma f x^2}{\Sigma f} - (\overline{x})^2}$

The variance has the same formulae but no square root signs, so:

For a list of numbers: Variance $= \dfrac{\Sigma x^2}{n} - \left(\dfrac{\Sigma x}{n}\right)^2$ or $\dfrac{\Sigma x^2}{n} - (\overline{x})^2$

For a frequency distribution: Variance $= \dfrac{\Sigma f x^2}{f} - \left(\dfrac{\Sigma f x}{f}\right)^2$ or $\dfrac{\Sigma f x^2}{f} - (\overline{x})^2$

In each case, use the second version if you have already calculated the mean, $(\overline{x})^2$.

In the following exercise, you may use your calculator but you must put down enough working so that your method can be followed. Use the appropriate formula from the introduction above.

If your answers are not exact, round them to three significant figures.

Exercise 8B

1 Calculate the standard deviation for each of these sets of data: **AO3**

  a 3 7 8 10

  b 2 5 9 12 13 16 17 20

  c 1.7 2.3 2.9 3.4 5.6 6.8 9.5 10.1 12.3 13.4

2 Jane's exam marks last summer were: **AO3**

  67 82 34 56 47 76 69 45 57

  Calculate:

  a her mean mark

  b the standard deviation of her marks

3 Calculate the standard deviation for the frequency tables below: **AO3**

  a

x	3	4	5	6
f	2	5	9	6

  b

x	4	6	8	10	12
f	3	12	8	5	2

  c

x	5	10	15	20	25	30
f	2	7	8	4	3	1

4 The table below shows the number of pets in 60 families: **AO3**

Number of pets	0	1	2	3	4	5
Frequency	4	17	21	9	8	1

  Calculate:

  a the mean

  b the variance

  c the standard deviation

5 The table below shows the number of televisions owned by 50 families: **AO3**

Number of TVs	0	1	2	3	4	5
Number of families	6	7	14	17	5	1

  Calculate:

  a the mean

  b the variance

  c the standard deviation

6 This table shows the number of books read this year by the 30 members of a reading group.

AO3

Number of books	Frequency
5	1
6	8
7	12
8	6
9	3

Calculate:

a the mean

b the variance

7 The weights of 50 dogs (to the nearest kg) have been collected and put into this grouped frequency table.

AO3

Weight (kg)	Number of dogs
1–3	5
4–6	7
7–9	11
10–12	16
13–15	11

Using the mid-point of each class as x, calculate:

a an estimate of the mean

b an estimate of the variance

8 Calculate an estimate of the standard deviation of each of these grouped frequency distributions, where x is a continuous variable:

AO

a

Class	Frequency
$1 \leqslant x < 5$	2
$5 \leqslant x < 9$	4
$9 \leqslant x < 13$	5
$13 \leqslant x < 17$	3
$17 \leqslant x \leqslant 21$	1

b

Class	Frequency
$10 \leqslant x < 20$	7
$20 \leqslant x < 30$	9
$30 \leqslant x < 40$	4
$40 \leqslant x < 50$	2
$50 \leqslant x \leqslant 60$	2

c

Class	Frequency
$0 \leqslant x < 5$	15
$5 \leqslant x < 10$	7
$10 \leqslant x < 15$	4
$15 \leqslant x < 20$	3
$20 \leqslant x \leqslant 25$	2

9 The heights of 80 seedlings have been measured correct to the nearest cm and the results put in this table.

AO3

Height (cm)	10–19	20–29	30–39	40–49	50–59	60–69	70–79
Number of seedlings	5	4	8	18	21	15	9

Calculate an estimate of the standard deviation.

10 For a set of data, $\Sigma f = 100$, $\Sigma fx = 347$ and $\Sigma fx^2 = 1942$.

AO3

Calculate the variance.

EQ 11 Stephanie recorded the time she took to travel to work on each of 50 days.

AO3

The table shows information about these times:

Time (x minutes)	Frequency (f)
$20 < x \leqslant 30$	4
$30 < x \leqslant 38$	9
$38 < x \leqslant 42$	12
$42 < x \leqslant 50$	18
$50 < x \leqslant 60$	7

a Calculate an estimate of the mean time Stephanie took to travel to work.

b Calculate an estimate of the standard deviation of these times.

You may use $\Sigma fx^2 = 91\ 367$.

12 Jada can sell all of the pumpkins grown on his farm to a supermarket, as long as 85% of them weigh between 750 g and 1.5 kg.

AO4

Weight, w (g)	$500 \leqslant w < 750$	$750 \leqslant w < 1000$	$1000 \leqslant w < 1250$	$1250 \leqslant w < 1500$	$1500 \leqslant w < 1750$
Frequency	22	63	98	53	7

Will the supermarket accept Jada's pumpkins?

8.3 Properties of frequency distributions

Quick reminder

A frequency distribution can be represented by a histogram but, if the widths of the classes get smaller and smaller, the histogram begins to look like a curve.

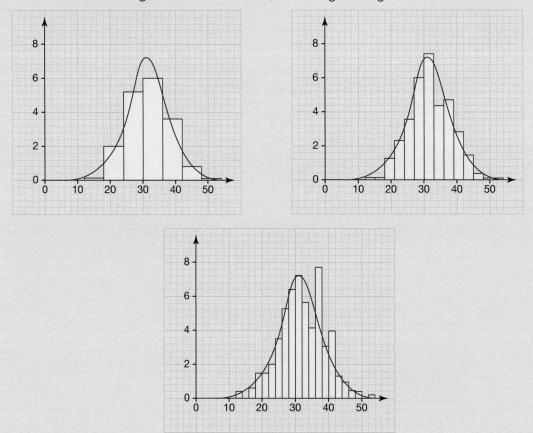

When we study the properties of frequency distributions, we usually represent them with curves.

Distributions can be **symmetrical** or **skewed**:

In a symmetrical distribution: median = mean = mode

If a distribution is positively skewed: median > mean > mode

If a distribution is negatively skewed: median < mean < mode

There are various ways to measure skewness, and you will be given the formula you need to use. The most common one is Pearson's which is:

$$\text{Skewness} = \frac{\text{mean} - \text{mode}}{\text{standard deviation}}$$

Quick reminder

A **normal distribution** is a bell-shaped symmetrical distribution that has many special properties. Nearly all of the distribution (99.8%) lies from 3 standard deviations below the mean to 3 standard deviations above.

e.g. if the mean is 50 and the standard deviation is 4, then:

$50 - 3 \times 4 = 38$ and $50 + 3 \times 4 = 62$, so nearly all the distribution lies between 38 and 62.

95% of the distribution lies between two standard deviations either side (between 42 and 58 in the example).

68% of the distribution lies between one standard deviation either side (between 46 and 54 in the example).

To compare exam marks for different subjects, we calculate **standardised scores**:

$$\text{Standard score} = \frac{\text{mark} - \text{mean}}{\text{standard deviation}}$$

The questions in this exercise incorporate material from both sections 8.3 and 8.4.

Exercise 8C

1 The frequency distribution graph below shows the cost of items in an art shop.

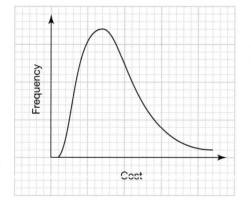

a Copy the distribution and indicate where the mean, median and mode would lie.   AO3

b Comment on the skewness of the distribution.   AO4

C

2 The two normal frequency distributions shown illustrate the times taken, in seconds, by two different groups of athletes to run 400 m.

Group A are javelin and hammer throwers.

Group B are 400 m runners.

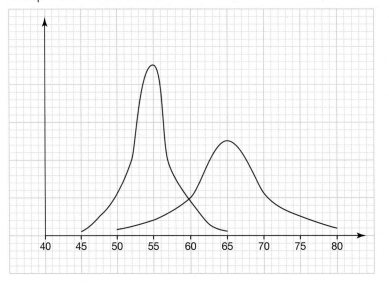

a Sketch the distributions and label them A and B.   AO3

b Comment on the respective abilities of both groups of athletes.   AO4

3 The following table gives the values of the mean, median, mode and standard deviation for five different distributions. Using the values of the mean, median and mode, say whether each distribution is symmetrical or positively skewed or negatively skewed.   AO

	Mean	Median	Mode	Standard deviation
a	23	23	23	5
b	23	28	32	18
c	23	19	15	12
d	15	15	15	7
e	35	28	21	10.5

4 Calculate Pearson's measure of skewness for each of the distributions in Question **3** using the formula: Skewness $= \dfrac{\text{mean} - \text{mode}}{\text{standard deviation}}$.   A

Give your answers correct to two decimal places where necessary.

5 The frequency distribution graph below shows the IQ scores of the 1350 students in a school.

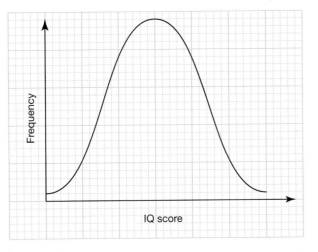

a Copy the frequency distribution and indicate where the mean, median and mode would lie. **AO4**

b Comment on the skewness of the distribution. **AO3**

c What is the name of this distribution? **AO3**

6 For each of the three frequency tables below calculate: **AO3**

i the mean

ii the mode

iii the standard deviation

iv the skewness, using Skewness = $\dfrac{\text{mean} - \text{mode}}{\text{standard deviation}}$.

Give your answers to three significant figures where necessary.

a

x	3	4	5	6
f	4	9	7	3

b

x	4	6	8	10	12
f	5	11	7	4	3

c

x	5	10	15	20	25	30
f	3	6	9	5	2	2

7 A normal distribution has mean 100 and standard deviation 20. **AO3**

Between what limits would you expect:

a 99.8% of the data to lie?

b 95% of the data to lie?

c 68% of the data to lie?

8 The weights, in kg, of 100 footballers and 100 rugby players are recorded. The results are normally distributed. The mean and standard deviation of each group is shown in the table below.

	Mean	Standard deviation
Footballers	76	10
Rugby players	85	15

   a Sketch the two frequency distributions on the same axis.   AO3

   b Make two comments about the distributions.   AO4

9 The length of time that customers queue at a supermarket checkout is normally distributed.   AO3

   The mean is 3 minutes and the standard deviation is 40 seconds.

   Between what limits would you expect 95% of the times to lie?

10 The time taken by a large number of pensioners to answer a problem follows a normal distribution with mean 75 seconds.   AO

   a If 95% of the pensioners took between 59 and 91 seconds, what is the standard deviation of the distribution?

   b Between what limits would you expect most of the times to lie?

11 A teacher calculates the mean and standard deviation of his class's maths exam results. He finds the mean is 62 and the standard deviation is 12. He uses these calculations to standardise the students' marks.   AC

   Calculate the standardised score for each of these marks:

   a 74   **b** 86   **c** 50

   d 44   **e** 62   **f** 65

12 The mean mark for physics in a teacher's class was 56. The standard deviation of the marks was 16. Calculate the actual mark scored if the standardised mark is: A

   a 1.5   **b** 2   **c** −1

   d 1.75   **e** −0.5

13 A veterinary nurse weighs each kitten born at the practice for one year. He finds that the weights are normally distributed with a mean weight of 87.5 g and a standard deviation of 18 g.   A

   What percentage of next year's kittens would you expect to have a weight of less than 33.5 g?

EQ 14 The height, in cm, of seedlings is normally a mean of 27 cm. It is found that 2.5% of the seedlings are at least 36 cm. Calculate the standard deviation of the heights of the seedlings.   A

9 Statistics used in everyday life

9.1 Statistics used in everyday life

Quick reminder

The Retail Price Index (RPI) measures how much the daily cost of living increases (or decreases).

One year is taken as the base year and given an index number, usually 100. Costs of goods and services in subsequent years are compared to those in the base year. Then each of these subsequent years is given a number, proportional to the base year, such as 103. An index of 103 means that the prices in that year are 103% of the prices in the base year, or that prices have increased by 3% from the base year.

Exercise 9A

1 The table shows the annual catch of fish in the Northeast Atlantic by UK trawlers. **AO3**

Year	1957	1967	1977	1987	1997	2007
Annual catch (1000 tonnes)	1009	922	979	911	886	594

   a Using 1957 as the base year, calculate a catch index for 1967, to 3 significant figures.

   b Using 1957 as the base year, calculate a catch index for 2007, to 3 significant figures.

   c What is the percentage decrease in catch from 1957 to 2007?

   d Using 1957 as the base year, 1953 had a catch index of 111. Calculate the annual catch of fish in 1953.

D

2 The graph shows the exchange rate for the Euro against the pound for each month in one year, 2009.

AO3

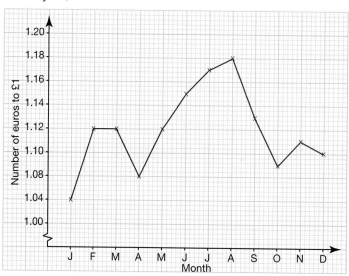

a What was the exchange rate in December 2009?

b Between which two months did the exchange rate rise the most?

c Explain why you could not use the graph to work out the exchange rate of December 2008.

AO

3 In 2005 the average cost of a bunch of flowers was £2.84. Using 2005 as a base year, the price index of the average cost of a bunch of flowers for each of the next five years is shown in the table below.

AO

Year	2005	2006	2007	2008	2009	2010
Index	100	112	132	144	135	142
Price	£2.84					

Copy and complete the table.

C

4 A manufacturer monitors the costs of a manufacturing process using weighted Index Numbers.

Type of cost	Weight	Index (2000 = 100)
Raw materials	82	206.5
Labour	185	150.1
Machinery and plant	120	128.6
Administration costs	65	34.8

The price index for each of the components in 2010 is known. The base year for the price index is 2000.

a Calculate the weighted index for the cost of the manufacturing process in 2010. Give your answer to 1 d.p.

b Which type of cost does the manufacturer see as the most important?

5 The table shows the annual cost of Annabella's car insurance for the past three years.

AO3

Year	2008	2009	2010
Annual cost	£248	£262	£321

a Use the chain base method to calculate index numbers for the years 2008 to 2010. Give your answers to one decimal place.

b What do these index numbers tell you about the percentage rise or fall in the cost of car insurance from 2008 to 2010?

c Give a possible reason for the largest of the percentage changes in the cost of Annabella's car insurance.

AO4

6 The Retail Price Index measures how much the daily cost of living increases or decreases.

AO4

If 2000 is given a base index number of 100, and 2010 is given 131, what does this mean?

7 The standardised death rate of a town is 8.3.

AO4

Explain what a standardised death rate of 8.3 means.

10 Time series and quality assurance

10.1 Time series and moving averages

Quick reminder

Time series analysis shows how the values of data change over time and can be used to identify patterns in the data. It can also be used to show underlying trends and seasonal variation.

Moving averages are used to smooth out seasonal variation and are calculated by grouping data, in sequence, depending upon the type of moving average that is required. For example, a three-point moving average would be obtained from the mean of the first three items of data, then the second to fourth items, and so on. A four-point moving average is based on the data, taken in fours.

Exercise 10A

1 This line graph shows the outside temperature at a weather station, taken at four-hour intervals during one day.

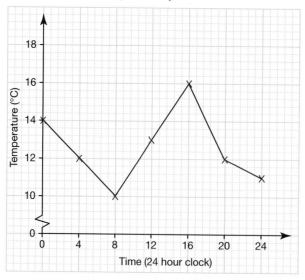

a At what time of the day was the lowest temperature?

b What was the highest temperature?

c Between which two readings was the greatest temperature change?

d Estimate the temperature at 10 am.

e When the outside temperature at the weather station goes above 13°C, the heating stops inside the building. For how long does the heating stop on this day?

2 Jon goes to a pub quiz once a week for eight weeks. The table shows his scores. **AO3**

Week	1	2	3	4	5	6	7	8
Score (out of 100)	35	47	54		68	74	78	82

a Draw a line graph for the data.

b Jon could not remember his score on the fourth week. Use your graph to estimate Jon's score on the fourth week.

c Explain the trend in Jon's scores. What reasons can you give to explain this trend? **AO4**

d Explain why it would be a mistake to use your graph to estimate Jon's score in the 12th week of the pub quiz. **AO4**

3 The table shows how much tourists spent in the UK between 1988 and 2008. **AO3**

Total spending (£ billions)	6.2	7.9	12.3	12.8	13.0	16.3
Year	1988	1992	1996	2000	2004	2008

a Draw a line graph for the data.

b Use your graph to estimate tourist spending in 2012. **AO4**

c Explain the trend in tourist spending. What reasons can you give to explain this trend? **AO4**

d Explain why many people think that tourist spending in 2012 will be much higher than your answer to part **b**. **AO4**

EQ 4 The time series graph shows information about the average number of barn owl eggs hatched in a different parts of a large protection area from 1997 to 2004.

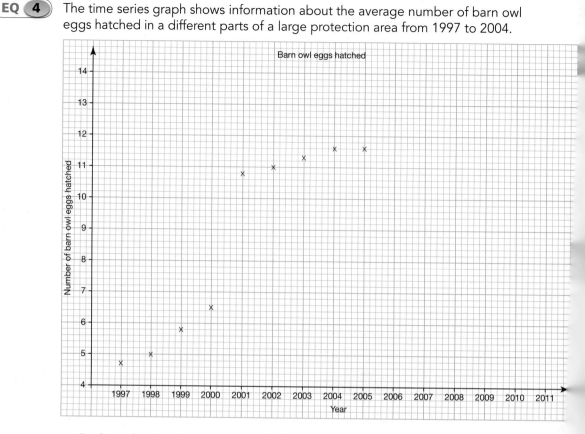

a Explain what the time series graph shows you about the average number of barn owl eggs that hatched in the protection area from 1997 to 2004.

b Use the information in the time series graph from 2001 to 2005 to predict the number of barn owl eggs that will hatch in 2011.

c Comment on the reliability of your prediction.

5 The table shows the number of goldfish born in a large garden pond from 2003 to 2010

Year	2003	2004	2005	2006	2007	2008	2009	2010
No. of goldfish born	66	68	60	51	44	36	30	25

a Plot a graph from the data in the table and draw a trend line for it.

b Use your graph to estimate the likely number of goldfish that will be born in 2011.

c Between which two years did the number of goldfish born decrease the most?

d What reasons can you give to explain this trend?

e Is it possible to use this data to predict the likely number of goldfish born in 2020?

6 The table shows the average price of a two-bedroom flat at certain distances from the seafront of a popular seaside town.

AO4

Distance from seafront (km)	0	1	2	3	4
Average price (£ thousands)	250	180	175	169	164

a Plot a graph from the data in the table and draw a trend line for it.

AO3

b Use your graph to estimate the average price of a two-bedroom flat 5 km from the seafront.

c What reasons can you give to explain this trend?

d Is it possible to use this data to predict the average price of a two-bedroom flat 10 km from the seafront? Explain your answer.

EQ 7 This table shows the cost of a household's gas consumption over a three-year period.

AO3

Year	2008				2009				2010			
Quarter	Q1	Q2	Q3	Q4	Q1	Q2	Q3	Q4	Q1	Q2	Q3	Q4
Cost (£)	134	78	52	120	144	76	59	133	158	84	60	
4-point moving average		96	98.50	98	99.75	103						

a Show that the first 4-point moving average is £96.

b Calculate the last three 4-point moving averages.

c Copy the graph below and plot the last three moving averages.

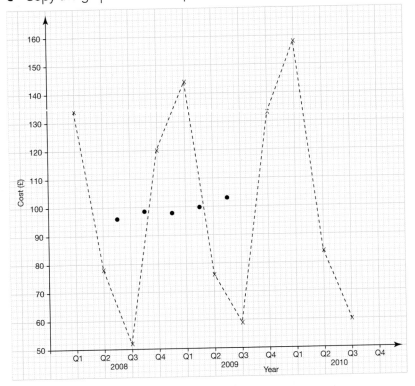

89

B

d Read off from the trend line the 'seasonal' effect on each 4th quarter.

e Use the trend of the moving averages to predict the cost of gas for the 4th quarter of 2010.

AO4

8 This table shows the number of calls, to the nearest 50, made to a call centre in 2010.

AO3

Month	Number of calls	Cumulative total for 2010	Calls over previous 12 months
January	1350	1350	18 750
February	1650	3000	19 000
March	1500		18 450
April	1200		17 600
May	1300		17 700
June	1350		17 500
July	1400		17 750
August	1250		17 300
September	1150		16 950
October	1250		16 600
November	1250		16 100
December	1250		15 900

a Copy and complete the column showing the cumulative total.

b Draw a Z chart for this data.

c What is happening to the number of calls to this call centre over time?

AO

A

EQ **9** A calculator manufacturer shows the following quarterly (Q), or three-monthly, sales figures (in thousands).

AO

Quarterly sales figures in 1000s			
	Q1	Q2	Q3	Q4
2010	42	46	40	42
2011	44	48	42	46

a Calculate the four-point moving averages for these data.

b Plot a graph showing the sales figures and the four-point moving averages.

c Comment on the trends seen in the sales of their calculators.

10.2 Quality assurance

Quick reminder

Quality assurance is very important to businesses, especially in manufacturing.

You can use sample means, medians and ranges to ensure consistency and accuracy.

Exercise 10B

1 Cement is put into bags which are supposed to contain 20 kg. A sample of 8 bags is taken and the mass of cement in each bag noted:

AO3

19.99 kg   20.01 kg   20.10 kg   19.95 kg   20.08 kg   20.25 kg   19.92 kg   19.70 kg

a Calculate the mean mass of the cement in the 8 bags sampled.

b Calculate the median mass of the cement in the 8 bags sampled.

c Are the median and mean helpful to the manufacturer? Explain your answer.   AO4

d An employee recommends taking another factor into account. What do you think this factor could be? Explain your answer.   AO4

2 Compost is put into bags which are supposed to contain 10 litres. A sample of 8 bags is taken and the capacity of compost in each bag noted:

10.75 l   10.73 l   9.89 l   9.91 l   9.61 l   10.77 l   9.85 l   10.09 l   AO3

a Calculate the mean capacity of the compost in the 8 bags sampled.

b Calculate the median capacity of the compost in the 8 bags sampled.

c Are the median and mean helpful to the manufacturer? Explain your answer.   AO4

d An employee recommends resetting the machines that fill the bags. Do you think this is a good idea? Explain your answer.   AO4

3 Cereal boxes are marked as containing 500 g of cereal.

A sample of ten boxes are checked and the mass of cereal in each box is:

502.4 g 500.8 g 499.7 g 503.0 g 502.1 g 499.6 g 501.2 g 502.4 g 502.5 g 502.1 g

a Calculate the mean mass of the cereal in the ten boxes.   AO3

b Does it appear that the manufacturer is putting enough cereal into each box? Explain your answer.   AO4

c By law, products must contain at least the amount printed on the packaging. Could any of the boxes pose a problem to the manufacturer? Explain your answer.   AO4

D

4 Aromatherapy oil is sold in bottles containing 30 ml of oil.

A trading standards officer buys five bottles of oil to test.

The volume of oil in each bottle is found to be:

31.1 ml   33.0 ml   30.4 ml   32.2 ml   29.9 ml

a Calculate the mean of the sample volumes.

AO3

b Do you think the bottles have too much oil in them? Explain your answer.

AO4

c What advice might the trading standards officer give to the company selling the bottles?

AO4

5 Chocolate boxes are marked as containing 400 g of chocolates.

A sample of six boxes are checked and the mass of chocolate in each box is:

402.6 g   400.8 g   398.7 g   403.0 g   402.1 g   400.6 g

a Calculate the mean mass of the chocolate in the six boxes.

AO3

b Does it appear that the manufacturer is putting enough chocolate into each box? Explain your answer.

AO4

c Could any of the six boxes pose a problem to the manufacturer? Explain your answer.

AO4

C

EQ 6 A machine is programmed to put 19 g of pepper into a container.

As part of quality assurance, a sample of 10 packets is taken every hour and the mean weight is calculated.

The results are shown in the following table:

Sample	Sample mean (grams, to 2 d.p.)
1	19.08
2	19.05
3	19.03
4	18.99
5	18.94

The quality control limits for this machine is: 19 g ± 1%.

a Work out the upper control limit.

b Are any of the sample means outside of the quality control limits for this machine?

c Do these sample means indicate that there are problems with the packing process?

d What action, if any, would you recommend that the machine programmer should take?

e The machine packs 8000 packets per hour. The 10 sample packets are taken together at the start of every hour. Is this a good method of sampling?

7 A manufacturer puts red chillies into jars which are supposed to contain 190 g. Six samples of ten jars are taken and the sample means calculated.

AO3

The results are shown in the table below:

Sample	Sample mean in grams
1	190.1
2	192.2
3	191.0
4	193.2
5	190.0
6	191.5

The quality control limit for the weight in the jars is 192 g ±1%.

a Work out the actual upper and lower quality control limits.

b Draw a graph with lines drawn in to represent the sample mean and the quality control limits (upper and lower).

c Plot the sample means on this graph.

d Do the results indicate that there is a problem with this production process? **AO4**

8 Toothpicks are packed into containers which should contain 150 toothpicks. Five samples of containers are taken and the sample mean and range calculated.

The results are:

Sample No.	Mean	Range
1	152	2
2	148	1
3	152	4
4	152	2
5	150	5

The acceptable range is 3 toothpicks.

a Copy the graphs below and plot the means and ranges on them. **AO3**

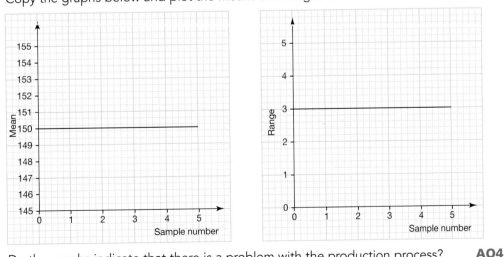

b Do the graphs indicate that there is a problem with the production process? **AO4**

c What should this manufacturer do? **AO4**

11 Correlation and regression

> ### Quick reminder
>
> A **scatter diagram** is a method of comparing two variables by plotting their corresponding values on a graph. These values are usually taken from a table.
>
> There are different types of **correlation** – positive, negative and no correlation.

Exercise 11A

1 Data is to be collected from students for each of the following pairs of variables. **AO**

Will this data be positively correlated, negatively correlated or have no correlation?

Give an explanation for each answer.

  a shoe size and height.

  b hair colour and number of friends.

  c mathematics set and science set.

  d hair colour and English set.

  e number of downloads bought and the amount of pocket money saved.

2 The heights and weights of 20 footballers are given in the table below. **A**

Height	1.61	1.73	1.69	1.75	1.79	1.95	1.76	1.80	1.66	1.7
Weight	67	70	71	74	73	85	73	78	65	77
Height	1.78	1.82	1.91	1.75	1.77	1.82	1.86	1.85	1.65	1.6
Weight	74	83	87	70	72	77	75	77	72	68

  a Draw a scatter diagram to illustrate the data.

  b Calculate the mean of both data sets.

  c Draw a line of best fit on the graph.

  d David is a 1.7 m tall footballer. How heavy would you expect him to be?

3 A health clinic counted the number of breaths per minute and the number of pulse beats per minute for ten people doing various activities. This information is shown in the table below.

AO3

Breaths per minute	16	20	20	24	26	28	28	30	34	36
Pulse beats per minute	58	68	70	72	84	80	84	88	94	102

a Draw a scatter diagram to illustrate the data.

b Calculate the mean of both data sets.

c Draw a line of best fit on the graph.

d Davila is breathing at a rate of 32 breaths per minute. Estimate her pulse beats per minute.

4 The table below shows the age and height of ten gifted young basketball players.   AO3

Age (years)	11.9	12.4	12.8	13.7	14.4	15.2	15.8	15.9	16	16.4
Height (metres)	1.52	1.44	1.64	1.84	1.77	1.85	1.81	1.88	1.95	1.90

a Plot the points on a scatter diagram.

b Calculate the means of age and height and use these to draw a line of best fit.

c Misca is a gifted basketball player who is 1.75 m tall. Estimate Misca's age.

d Explain why it would not be sensible to extend the line of best fit so that you could determine the height of a 20-year-old basketball player.   AO4

5 The table below shows the test marks for ten students in their English and history tests.

AO3

Student	Ali	Brian	Clare	Dai	Ed	Farr	Greg	Henry	Ira	Jamir
English	35	61	62	27	34	59	66	45	82	39
History	36	35	54	25	30	55	60	45	75	65

a Plot the data on a scatter diagram. Take the x-axis for English and the y-axis for history.

b Calculate the mean scores for English and history and use these to draw a line of best fit.

c One student felt unwell on the day of the English test. Who was this person? Explain your answer.   AO4

d One student felt unwell on the day of the history test. Who was this person? Explain your answer.   AO4

e Kali scored 70 marks in her English test but was absent for the history test. Estimate what score Kali would have got if she had taken the history test.   AO4

6 A tyre repair workshop has a faulty tyre pressure gauge.   AO3

The gauge is checked against a correct gauge and the following data is obtained:

Reading of gauge (bar)	1.0	1.4	1.8	2.2	2.6	3.0	3.4	3.8
Correct reading (bar)	1.96	2.23	2.85	3.19	3.58	4.02	4.33	4.70

a Plot the points on a scatter diagram.

b Calculate the means both of the reading of the gauge and of the correct reading.

c Draw a line of best fit on your graph.

d Use your graph to work out the equation of the line in the form $y = mx + c$.

e What is your interpretation of where the line crosses the y-axis?　　AO4

7 The data from an experiment comparing reaction time for a simple task and the blood/alcohol concentration (BAC) is given in the table below:　　AO3

BAC (g/l)	0.01	0.02	0.05	0.08	0.10	0.12	0.18
Reaction time (seconds)	1.5	1.9	3.0	5.2	6.4	7.2	12.6

a Plot the points on a graph.

b Calculate the means of both blood/alcohol concentration and reaction time.

c Draw a line of best fit on your graph.

d Use your graph to work out the equation of the line in the form $y = mx + c$.

e What is your interpretation of where the line crosses the y-axis?　　AO

11.2 Spearman's rank and product moment correlation coefficient.

Quick reminder

Spearman's rank correlation coefficient (SRCC) is used:

- when only rankings are given, for example, when two wine tasters each put 10 wines in order of preference

- if you think there is a non-linear relationship in the data: for example, a scatter diagram that forms a curve, such as the correlation between the number of hours an athlete trains and the improvement in the athlete's performance. You would imagine that it would increase steadily up to a point and then start to taper off into a curve when excessive training does not necessarily improve performance.

The formula to calculate Spearman's rank correlation coefficient is:

$$\text{SRCC} = 1 - \frac{6\Sigma d^2}{n(n^2 - 1)}$$

where n is the number of data values

Σd^2 is the sum of the differences between the ranks squared.

The product moment correlation coefficient (PMCC) is used when a scatter diagram shows an approximately linear relationship, i.e. a straight line.

Exercise 11B

1 Michelle calculates Spearman's rank correlation coefficient for three different sets of data. Her results are shown in the table below: **AO4**

Data sets	Spearman's rank correlation coefficient
a Hours of sunshine vs. amount of rain	−0.92
b Size of diamond vs. price of diamond	0.89
c Eye colour vs. height	0.21

Comment on the relationship for each data set.

EQ 2 The following table shows the population and land area of eight European countries.

Country	Land area (km²)	Population (millions)	Land area rank	Population rank	d	d^2
Austria	83 871	8.21				
Belgium	30 528	10.42				
France	643 427	64.77				
Germany	357 022	82.28				
Luxemburg	2 586	0.49				
Ireland	70 273	4.62				
Spain	505 370	46.51				
UK	243 610	62.35				

a Copy and complete the table.

b Calculate Spearman's rank correlation coefficient for these data.

c Interpret your answer to part **b** in full.

3 The table below shows the personal best times of ten athletes for running 5000 m and 10 000 m in minutes and seconds.

AO3

Athlete	5000 m	10 000 m	5000 m rank	10 000 m rank	Difference d	d^2
A	15.36	31.34				
B	16.55	32.33				
C	13.50	30.02				
D	14.25	29.44				
E	17.32	38.23				
F	20.12	36.40				
G	14.48	29.57				
H	15.45	30.09				
I	14.58	28.52				
J	16.04	32.35				

a Copy and complete the table.

b Use the table to calculate Spearman's rank correlation coefficient.

c Comment on the relationship between the two sets of data.

AO

4 The table below shows the average daily rainfall and the average number of hours of sunshine at a weather station.

AO

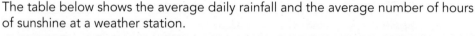

Month	Rainfall (mm)	Sunshine (hours)	Rainfall rank	Sunshine rank	Difference d	d^2
January	1.36	1.2				
February	1.35	2.7				
March	0.75	4.6				
April	2.22	5.2				
May	2.54	5.7				
June	2.26	7.7				
July	2.99	5.2				
August	1.84	5.8				
September	2.66	4.9				
October	1.74	3.1				
November	1.57	2.9				
December	2.09	1.9				

a Copy and complete the table.

b Use the table to calculate Spearman's rank correlation coefficient.

c Comment on the relationship between the two sets of data.

5 Mr Meaner and Ms Take are judges at an art competition. The best ten paintings are ranked by each of them. The results are in the table below:

Painting	A	B	C	D	E	F	G	H	I	J
Mr Meaner	5	3	10	5	7	8	1	2	9	4
Ms Take	8	1	9	7	4	5	3	2	10	6

   a Use the table to calculate Spearman's rank correlation coefficient.   **AO3**

   b Comment on how well (or not) the two judges agree.   **AO4**

6 Alicia thinks that her friends who are good at Spanish are also good at German. Is she correct?   **AO4**

Name	Zola	Yves	Xavier	Will	Viki	Uri	Toni	Sal	Rico	Qamra
Spanish exam result	9	12	16	12	10	7	12	12	8	16
German exam result	10	10	13	8	6	4	7	6	5	7

7 Using anonymous data from her local hospital, Lily draws up the following table: **AO4**

Patient	A	B	C	D	E	F	G	H	I	J
Number of years smoking	12	23	24	28	30	33	37	42	43	47
Percentage damage to lungs	18	52	55	28	59	41	59	66	61	69

Lily thinks that the longer a person smokes, the worse their lung damage is likely to be. Is she correct?

12 Probability

12.1 Probability scale

> **Quick reminder**
>
> **Probability** is the chance or likelihood that something will happen. All probabilities lie between 0 and 1.
>
> A probability of 0 means it is **impossible** for the event to happen.
>
> A probability of 1 means the event is **certain** to happen.

Exercise 12A

G

 Here are some probability values.

1   0.5   0.25   0.8   0

Match the probabilities with the words in the table.

Word	Probability
Even chance	
Certain	
Likely	
Impossible	
Unlikely	

 Rewrite these words in order of increasing probability.

likely   impossible   very likely   very unlikely   even chance   certain

EQ 3 Give the value for the probability of an event that is:

a certain to happen

b impossible

c equally likely to happen or not happen.

 Use suitable probability words to describe events that have a probability of

a 0.9

b 0.12

5 Match each of these words to each statement.

impossible   certain even chance   likely   unlikely

a I will wake up in the morning with seven heads.

b I will get tails when I throw a fair coin.

c I will be older next week than I am today.

d I will watch television this week.

e I will have no homework during a school week.

EQ 6 A fair six-sided dice is rolled. Events P, Q and R defined below.

Event P: the dice shows an odd number.

Event Q: the dice shows the number 7.

Event R: the dice shows the number 4 or less.

a Copy and complete the labelling of the probability scale.

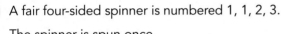

0   ....

b Put arrows on the probability scale to show the probability of each event. Label the letters with the correct letter P, Q or R.

7 A fair four-sided spinner is numbered 1, 1, 2, 3.

The spinner is spun once.

Copy the probability scale. Put an arrow on the scale to show the probability of each of these outcomes.

0   0.5   1

a The spinner lands on a 1. Label this A.

b The spinner lands on a number greater than zero. Label this B.

c The spinner lands on an odd number. Label this C.

8 Ali says, 'It might rain today, or it might not rain today. So the probability of it raining today is an even chance.'

AO3

Explain why Ali might be wrong.

12.2 Equally likely outcomes

Quick reminder

The possible results of a trial are called **outcomes.**

The probability of an event happening = $\dfrac{\text{number of successful outcomes}}{\text{total number of possible outcomes}}$

This is known as **theoretical probability**.

The probability of an event **not** happening is 1 – the probability of the event happening.

Exercise 12B

1 A pencil case contains seven red pens and four black pens. If a pen is taken out of the pencil case at random, what is the probability the pen is

   a a red pen

   b a blue pen

   c a black pen?

2 A box of chocolates contains 15 chocolates with hard centres and 13 chocolates with soft centres. If a chocolate is taken out of the bag at random, what is the probability that it will be:

   a a hard centre

   b a soft centre

   c a biscuit?

EQ 3 A bag contains three red balls, four blue balls and six green balls. If a ball is chosen at random from the bag, what is the probability that it is:

   a a red ball

   b a green ball

   c a yellow ball

   d a blue or red ball?

EQ 4 A box contains 21 counters: 7 are red, 4 are blue and the rest are green. If a counter is taken at random from the box, what is the probability that it is:

   a red

   b green

   c red or green?

5 One letter is chosen at random from the word MATHEMATICS.

Work out the probability that the letter chosen is:

a the letter S

b the letter M

c a vowel.

6 A card is drawn at random from a full pack of 52 playing cards. What is:

a P(four of Hearts)

b P(a Club)

c P(a black card)

d P(a Queen)

e P(not a Queen)?

7 In a raffle, 1000 tickets are sold. Amy buys 12 tickets, Ruby buys 3 tickets, Colin buys 20 tickets and Rebecca buys 6 tickets. If there is only one winning ticket, what is the probability that:

a Ruby wins

b Colin wins

c one of the girls wins

d Amy will not win?

8 The probability of picking an ace from a pack of cards is $\frac{1}{13}$.

What is the probability of **not** picking an ace from a pack of cards?

9 The probability of Dai being late to his lesson is 0.6.

What is the probability of Dai **not** being late to the lesson?

10 The probability that Alan is late for work is 0.2.

Work out the probability that Alan is **not** late for work.

11 The probability that Sarah sends a text on any given day is 0.72.

Work out the probability that Sarah does **not** send a text on any given day.

12 The probability that Rhydian buys a chocolate bar is $\frac{7}{11}$.

What is the probability that Rhydian does **not** buy a chocolate bar?

12.3 The addition rule for events

Quick reminder

Mutually exclusive events are events that cannot happen at the same time, for example throwing an even number on a dice and throwing an odd number on a dice. Since there are no other possibilities, they are also called exhaustive events.

For mutually exclusive events

P (A or B or C or ....) = P(A) + P(B) + P(C) + ...

Exercise 12C

D

1 A box of chocolates contains 13 chocolates which look identical from the outside. Five of the chocolates have toffee centres, seven have soft centres and one has a nut centre.

One chocolate is taken at random from the box.

What is the probability that the chocolate:

a has a soft centre

b has a nut centre

c has a toffee or nut centre

d has a soft or toffee centre

e does not have a soft centre?

EQ 2 A bag contains three red beads, four blue beads and five green beads. A bead is taken from the bag at random. What is the probability of choosing:

a a green bead

b a blue bead

c a green or red bead

d a red, blue or green bead?

3 A letter is chosen at random from the word STATISTICS. What is the probability that it is:

a the letter S

b the letter I

c the letter A or T

d a vowel

e the letters S, A or T?

4 Libby takes a card at random from a full pack of 52 playing cards. What is:

   a P(three of Spades)

   b P(a black card)

   c P(a Heart)

   d P(a Heart or the three of Spades)

   e P(a red card or the three of Spades)?

5 James has a set of 11 cards numbered from 1 to 11.

   He picks out a card and then replaces it in the pack.

   What is the probability that the number on the card is:

   a a multiple of 3

   b an odd number

   c an even number greater than 4

   d an odd number or an even number greater than 4?

6 Sarah, Lucy and Josh are playing a card game. The probability that Sarah wins is 0.3 and the probability that Josh wins is 0.25. Who is more likely to win the game?

7 Rob, Jack and Will are playing darts. The probability that Rob wins is 2/7. The probability that either Rob or Will win is 3/5. Which of the three players is most likely to win?

8 Broughton, the famous pub quizzer, has a random question generator on his iPad. It which contains 16 000 pub-quiz questions.

   AO3

   It has four basic categories: 'science and nature', 'history and geography', 'literature, art and music' and 'weird stuff'.

   The probability of a 'history and geography' or a 'literature, art and music' question coming up is 0.55.

   The probability of a 'history and geography' or a 'science and nature' question coming up is 0.65.

   The probability of a 'literature, art and music' question **not** coming up is 0.7.

   Use this information to copy and complete the table below:

Type	Probability	Number of questions
Science and nature		
History and geography		
Literature, art and music		
Weird stuff		

12.4 Experimental probability

Quick reminder

In real-life situations, the probability of different outcomes are not always equal or possible to work out. In such cases an experiment will need to be carried out. This is called experimental probability or relative frequency. The relative frequency of an event is an estimate for the theoretical probability.

Relative frequency of an outcome or event = $\dfrac{\text{the number of successful outcomes}}{\text{the total number of trials}}$

Exercise 12D

1 You can find probabilities by: AO4

  A looking at historical data

  B using equally likely outcomes

  C using a survey or experiment.

Which method would you use to find the probability that:

a when a dice is rolled, the number will be a 1

b Adrian would be good at taking penalty kicks

c Sam will be late to school some time next week

d a three-sided spinner is fair?

2 Thomas throws a fair six-sided dice and records how many times he gets a two.

After 50 throws, he has scored 4 twos.

After 100 throws, he has scored 9 twos.

After 150 throws, he has scored 16 twos.

After 200 throws, he has scored 21 twos.

After 600 throws, he has scored 91 twos.

After 1000 throws, he has scored 168 twos.

After 2000 throws, he has scored 331 twos.

  a What is the theoretical probability of throwing a two with a dice?

b Calculate the experimental probability of scoring a two after:

   i 50 throws

   ii 100 throws

   iii 150 throws

   iv 200 throws

   v 600 throws

   vi 1000 throws

   vii 2000 throws.

c If Thomas threw the dice 12 000 times, how many twos would you expect him to get?

3 Jessica carries out an experiment to work out the probability that when she drops a drawing pin, it will land point up.

The table shows her results after 100, 200, 500, 1000, 1500, 2000, 2500 and 3000 trials.

Number of times pin is dropped	100	200	500	1000	1500	2000	2500	3000
Number of times pin lands point up	87	148	335	584	883	1182	1492	1797

a Calculate the experimental probability of a drawing pin landing point up after

   i 100 trials

   ii 200 trials

   iii 500 trials

   iv 1000 trials

   v 1500 trials

   vi 2000 trials

   vii 2500 trials

   viii 3000 trials.

b What do you think the experimental probability of the drawing pin landing point up is?

c If 18 000 of these drawing pins were dropped, how many would you expect to land points up?

4 Verity works in a factory making chips for computers. One of her jobs is to test a sample each day to make sure they work properly.

DAY	Monday	Tuesday	Wednesday	Thursday	Friday
Number tested	910	670	1250	887	526
Number faulty	13	9	19	11	5

On which day is it most likely that the highest number of faulty computer chips were produced?

C

5 400 drivers in Manchester were asked if they had ever gone down a road with a 'No Entry' sign. 27 answered that they had.

There are 148 000 drivers in Manchester. How many of these do you estimate will have gone down a road with a 'No Entry' sign?

6 Rupinder spun this spinner 300 times.

She said, 'If this is a fair spinner I will get 100 number 4s.'

Explain why she is wrong.

7 Jason takes his young son Aaron to a 'ball pool'.

AO3

A sign says there are 500 000 coloured balls in the pool.

Aaron throws balls to Jason at random. Jason notes their colours and throws them back.

Colour	Red	Blue	Yellow	Green
Frequency	37	19	29	15

Work out how many red balls there are likely to be in the ball pool.

8 Samuel throws a six-sided dice 120 times with these results.

Number on dice	1	2	3	4	5	6
Frequency	17	19	11	33	22	18

a Why might Samuel think that the dice is biased?

b What could Samuel do to make his results more reliable?

EQ **9** Kuschal and Emily are testing a spinner to see if it is fair. The spinner has four equal sections: one is blue, one is green, one is black and one is red.

Kuschal spins the spinner 20 times it lands on black 4 times.

a What is the relative frequency of it landing on black after these 20 spins?

b Emily spins the spinner 100 times it lands on black 23 times. Explain why Emily's results should be more reliable than Kuschal's results.

c Combine Kuschal's data and Emily's data to estimate the probability that this spinner lands on black.

10 Which of these would suggest bias?

a Getting a Tail 2 times when flipping a coin 15 times.

b Getting 23 ones on a four-sided spinner labelled 1 to 4, in 100 spins.

c Getting 3 fives when rolling a dice 60 times.

12.5 Combined events

Quick reminder

When two events happen at the same time, all possible outcomes can be shown in a **sample space diagram.** One event is written in the rows of the table and one event is written in the columns of the table.

ercise 12E

1 A coin and fair six-sided dice are thrown. The sample space diagram shows the possible outcomes.

		Dice					
		1	2	3	4	5	6
Coin	Head	(H, 1)	(H, 2)	(H, 3)	(H, 4)	(H, 5)	(H, 6)
	Tail	(T, 1)	(T, 2)	(T, 3)	(T, 4)	(T, 5)	(T, 6)

What is the probability of getting a tail and a two?

EQ 2 Two four-sided spinners are used in a game. The first spinner is labelled 2, 4, 6, 8 and the second spinner is labelled 3, 5, 5, 7.

Both spinners are spun and their scores added together.

This sample space diagram shows all the ways the two spinners can land.

		Spinner 1			
		2	4	6	8
Spinner 2	3				
	5				
	5				
	7				

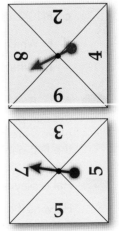

a Copy and complete the table to show the total scores.

b When the two spinners are spun together, what is the probability that the total score will be:

i an even number

ii 11

iii 7 or 13?

EQ 3 Two fair dice are used in a game.

Both dice are thrown and their scores **multiplied** together.

a Draw a sample space diagram to show the total scores.

b When the two dice are thrown together, what is the probability that the total score will be:

   i an odd number

   ii a score of 7

   iii a score of 12

   iv a multiple of 5?

4 There are two spinners: one with 3 sides numbered 2, 3, and 4 and the other with 8 sides numbered 1, 2, 3, 4, 5, 6, 7 and 8. The two spinners are spun and their scores are added together. What is the probability that the sum is:

a 7

b odd

c less than 6?

5 A three-sided spinner numbered 2, 2 and 3 and a four-sided dice, numbered 1, 2, 3 and 4 are thrown. The two scores are **multiplied** together.

What is:

a P(2)

b P(9)

c P(10)?

d i Which is the most likely score?

   ii What is the probability of that score?

6 Alice, Beth, Craig and Daniel are playing in a squash competition. Each player in the competition plays every other player. There are six matches altogether. Two players are picked at random to play the first game. Work out the probability that the first game will be played by a male player and a female player.

7 Hal is playing a game with two fair spinners.

He wins if both spinners land on 'Win'.

What is the probability of Hal winning?

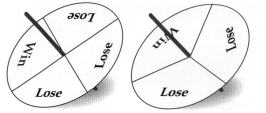

8 Ishmael makes this game to raise money for charity.

A03

Contestants pay £1 to play.

They roll two fair 10-sided dice, each numbered from 1 to 10.

If they score a 5 on one dice, they win £1.

If they score a total of 5, they win £2.

If they score a double 5, they win £10.

Is he likely to make or lose money for the charity?

12.6 Expectation

Quick reminder

When you know the probability of an event, you can predict how many times that you would expect the event to happen. This is called **expectation.**

Exercise 12F

1 The probability that Alan is late to work is 0.2. Alan works for 40 days before his next holiday. On how many days would you expect Alan to be late for work?

2 The probability that Sarah sends a text on any given day is 0.7. Work out how many texts (to the nearest whole number) that you would expect Sarah to send in one week (7 days).

3 A bag has some blue counters and some red counters. The probability of choosing a blue counter is 0.65. There are 500 counters in the bag. How many counters are blue?

 4 Varia has a bag containing four red balls, two green balls and three yellow balls.

She chooses one at random, notes the colour and puts it back.

She does this 81 times.

How many of each colour should she expect to get?

5 The probability that it rains on any day in April is 0.6.

How many rainy days would you expect in April?

EQ 6 A five-sided spinner is biased and has these probabilities of landing on each face.

Number	1	2	3	4	5
Probability	0.2	0.1	0.3	0.25	

a What is the probability that it lands on 5?

b I spin the spinner 200 times, how many times might I expect it to land on 3?

7 Amy buys a scratch card. The probability she wins a prize is 2/7.

   a What is the probability that Amy does not win a prize?

   b Amy buys 21 scratch cards. On how many cards would you expect Amy to win a prize?

8 The probability that a light bulb is faulty is 0.04. In a box of 400 light bulbs, how many would you expect to be faulty?

EQ 9 A school snack bar offers a choice of four snacks: fish, pizza, pasta and chicken.

Students can choose **one** of these four snacks.

The table shows the probability that a student will choose burger or pizza or chicken.

Snack	Fish	Pizza	Pasta	Chicken
Probability	0.25	0.35		0.1

600 students used the snack bar on Wednesday. Work out an estimate for the number of students who chose pasta?

EQ 10 A bag contains some yellow balls, some blue balls, some green balls and some pink balls.

William is going to take one ball at random from the bag.

The table shows the probabilities that the ball will be yellow, blue or pink.

Colour	Yellow	Blue	Green	Pink
Probability	0.34	0.17		0.28

William repeats the experiment 200 times, replacing the ball each time.

Work out an estimate for the number of green balls William selected.

11 Alex buys five raffle tickets. She is told that her probability of winning is 1/125.

How many tickets are there in the raffle altogether?

12 Katrin makes this game to raise money for charity.

Contestants pay £1 to play.

They roll two fair 10-sided dice, each numbered from 1 to 10.

If they score a 10 on one dice, they win £1.

If they score a total of 10, they win £5.

If they score a double 10, they win £10.

100 people play the game.

Is Katrin likely to make or lose money for the charity? How much?

AO

13 If I roll two fair normal dice 500 times, on how many occasions would I expect to score a total of 5 or less?

A

12.7 Tree diagrams

Tree diagrams are useful when working out the probability of combined events.

Exercise 12G

 1 Bina has two fair four-sided spinners, one red and one yellow.

On the red spinner there are three triangles and one square, and on the yellow spinner there are two triangles and two squares.

Bina spins each spinner once.

Copy and complete the tree diagram to show the probabilities when each spinner is spun.

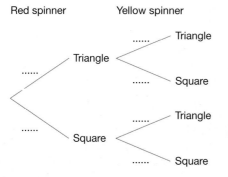

a What is the probability that both spinners land on triangles?

b What is the probability that one of the spinners lands on a triangle and the other spinner lands on a square?

B

EQ 2 Chloe throws a fair red dice once and a fair green dice once.

a Copy and complete the probability tree diagram to show the outcomes. Label clearly the branches of the probability tree diagram.

The probability tree diagram has been started in the space below:

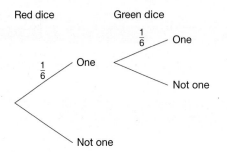

b What is the probability that both spinners land on a one?

c What is the probability that only one of the spinners lands on a one?

EQ 3 Jonathan plays one game of tennis and one game of darts.

The probability that Jonathan will win at tennis is $\frac{4}{7}$.

The probability that Jonathan will win at darts is $\frac{1}{3}$.

a Copy and complete the probability tree diagram.

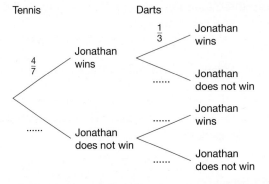

b Work out the probability that Jonathan wins both games.

c Work out the probability that Jonathan wins just one game.

EQ 4 The probability that Matthew is late for school on any particular day is 0.4. The probability that Daisy is late for school on any particular day is 0.65.

The probabilities of Matthew and Daisy being late for school are independent.

a Copy and complete the probability tree diagram.

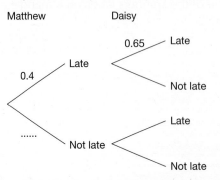

b Work out the probability that both Matthew and Daisy will be late for school on any particular day.

c Work out the probability that Matthew will be late and Daisy will not be late for school on any particular day.

d Work out the probability that either Matthew or Daisy will be late for school on any particular day.

EQ 5 Ben travels to work on a bus and a train. The probability that the bus is on time is 0.7. The probability that the train is on time is 0.6. Both events are independent of each other.

a Copy and complete the probability tree diagram.

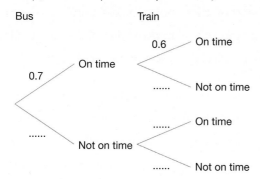

Bus   Train

0.7 On time
     0.6 On time
     ...... Not on time

...... Not on time
     ...... On time
     ...... Not on time

b Work out the probability that both the bus and the train are on time.

c Work out the probability that the bus is on time and the train late.

d Work out the probability that both the bus and train are late.

EQ 6 Mr Smith and Mrs Tate both go to the library every Wednesday. The probability that Mr Smith takes out a fiction book is 0.8. The probability that Mrs Tate takes out a fiction book is 0.4. The events are independent.

a Copy and complete the probability tree diagram.

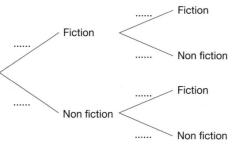

Mr Smith   Mrs Tate

...... Fiction
     ...... Fiction
     ...... Non fiction

...... Non fiction
     ...... Fiction
     ...... Non fiction

b Calculate the probability that both Mr Smith and Mrs Tate take out a fiction book.

c Calculate the probability that only one fiction book is taken out.

7 Zi Ying and Katie go to the canteen for lunch. Main meals are served with either mashed potato or jacket potato.

The probability that Zi Ying chooses mashed potato is 0.4.

The probability that Katie chooses mashed potato is 0.7.

Draw a probability tree diagram and use it to calculate the probability that:

a they both choose mashed potato

b they both choose jacket potato

c one chooses mashed potato and the other chooses jacket potato?

8 A pencil case contains 12 pens that look identical but that have different coloured ink.

Eight pens have blue ink, four have black ink.

Kate takes a pen, then Richard takes a pen.

a What is the probability that they both choose a blue pen?

b What is the probability that they both choose a pen of the same colour?

c What is the probability that they choose one of each colour?

EQ 9 Shamil, Charlotte and Reece are taking their driving tests.

The probability that Shamil passes his driving test on the first attempt is 0.5.

The probability that Charlotte passes her driving test on the first attempt is 0.7.

The probability that Reece passes his driving test on the first attempt is 0.4.

Draw a probability tree diagram and use it to calculate the probability that:

a all three will pass the driving test on the first attempt

b all three will fail the driving test on the first attempt

c Charlotte passes the driving test and Shamil and Reece both fail.

10 The probability that Sally is late for work is 4/7. The probability that Anil is late for work is 1/5. The probability that Richard is late for work is 1/2.

Draw a probability tree diagram and use it to calculate the probability that on a particular day that:

a all three are late

b none of them are late

c at least one of them is late?

11 Lucy takes four A-Levels.

The probability that she will pass mathematics is 0.7.

The probability that she will pass geography is 0.6.

The probability that she will pass biology is 0.85.

The probability she will pass chemistry is 0.95.

Draw a probability tree diagram and use it to calculate the probability that she passes

a all four subjects

b exactly two subjects

c at least two subjects?

12.8 Conditional probability

Quick reminder

The term **conditional probability** is used to describe a situation where the probability of an event is dependent on the outcome of another event.

Exercise 12H

1 Mrs Harris drives down a country lane on her way to work .The probability that she meets a tractor in the lane is 0.2. If she meets a tractor the probability that she is late for work is 0.6. If she does not meet a tractor the probability that she is late for work is 0.1.

   a Copy and complete the probability tree diagram.

   b What is the probability that Mrs Harris meets a tractor and is late for work?

   c What is the probability that Mrs Harris is not late for work?

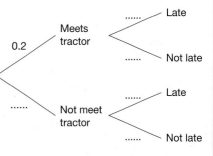

EQ 2 The head teacher of a school is concerned about the returns of books to the library. He carried out a survey of the students from the lower school, the upper school and the sixth form to see if they returned library books early (E), on time (T) or late (L).

Of the students in the lower school, 26% returned the books early, 70% returned the books on time and the remainder returned them late.

Of the students in the upper school, 10% returned the books early, 58% returned the books on time and the remainder returned them late.

Of the students in the sixth form ,18% returned books early, 80% returned them on time and the remainder returned them late.

   a Use this information to copy and complete the tree diagram.

   b A student is chosen at random from the school database. Work out the probability that they handed in their library book either early or on time.

   c Another student is chosen at random from the school database. The student is in the upper school. Calculate the probability that the student handed their library book in early or on time.

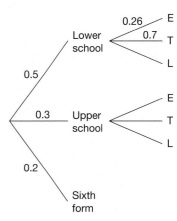

A*

EQ 3 A bag has three green and seven red counters. Two counters are taken at random from the bag, without being replaced.

    a What is the probability that both counters are green?

    b What is the probability that one counter is red and the other counter is green?

EQ 4 A tin contains 15 ginger biscuits and 12 lemon biscuits. Florence takes two biscuits at random.

    a What is the probability that both biscuits are ginger?

    b What is the probability that Florence has a ginger biscuit and a lemon biscuit?

EQ 5 The probability that it will be fine tomorrow is 0.6. If it is fine tomorrow, the probability that it will be fine the day after is 0.85. If it is not fine tomorrow, the probability it is fine the next day is 0.2.

What is the probability that it will be:

    a fine both tomorrow and the next day

    b fine on at least one of the two days?

EQ 6 A driving school knows from experience that the probability of one of its students passing the driving test at the first attempt is 0.6. If a student fails a test at any attempt, the probability of them passing next time is 0.75

What is the probability that a student at this driving school has not yet passed the test after three attempts?

EQ 7 A bag contains six red counters, three blue counters and two yellow counters. Two counters are taken out of the bag at random and without replacement.

    a What is the probability of taking a blue counter followed by a red counter?

    b What is the probability that both counters are the same colour?

EQ 8 A test is in two parts: a written test and a practical test. It is known that 85% of those who take the written test pass. When a person passes the written test, the probability they will pass the practical test is 50%. When a person fails the written test, the probability that they will pass the practical test is 12%.

What is the probability that someone:

    a passes both tests

    b fails both tests

    c passes just one test?

What can you do with Statistics?

Real-life Statistics

Discover exciting careers that use data analysis

How are Statistics used by ...

... Software developers?

The job

Software is everywhere. There is intricate programming and development behind all the millions of Apps and programs you use every day. Developers create everything from new versions of classic programs on your PC to the addictive games on your mobile. They are involved in designing and writing (coding) software, so that you can use everything as easily as possible.

You could be:
- A software engineer
- A software publisher
- A computer programmer
- A web developer
- A systems analyst
- A solution engineer

The maths

Software engineers learn a complex coding language which they use daily to tell the computer how to act when a software user performs a certain action. In developing the programs, the programmers use trial and error, optimisation, data organisation, statistical analysis, modelling and market research to ensure their intended audience is pleased with the program navigation.

The profile

George is a Game Developer at a busy video, computer and mobile gaming company. He was excited to hear about a career that combined his degree in Applied Maths and his love of video games. On a daily basis, he is developing, de-bugging, testing and playing games that go out in all different formats. Because of his natural interest in mobile phone and tablet application development, he has recently taken on a project to create App versions of his company's top five video games.

GIVE IT A TRY!

1. The table below gives the amount of hours spent by each department when producing a new game. Create a pie chart to display the information.

Software engineer	150
Software publisher	950
Computer programmer	2400
Web developer	300
Systems analyst	1000

2. George has conducted a survey about the types of games his 50 friends play. There are three types of game; strategy, simulation and platform.

   In total, 40 people like to play strategy games. Of these 40, 21 also play simulation games and 5 play platform games as well as strategy games. Nobody liked only platform games.

   If 3 people like to play all three types of game and 4 people like only simulation games, how many people play simulation and platform games? Use the information to create a Venn diagram to represent the information.

3. George also wants to find out how much time people spend playing computer games each week. He goes along to a games shop and asks shoppers there to complete his survey. Explain why his results may be biased. Suggest how he may improve his results.

… Business consultants?

The job

Businesses all over the world look to experts when it comes to responsible use of money, time and leadership. These professionals can either be in-house or part of an outside firm and are charged with assessing a company's various components to find trouble spots or ways to save money.

You could be:
- An actuary (Insurance Statistician)
- A risk management consultant
- An efficiency consultant
- A management consultant

The maths

A business consultant looks a lot at past figures to get a picture of a company's daily activity. They will often run surveys or do sampling within departments to get feedback from employees and/or customers. They look at percentages, too, aiming for low risk, high profits, high efficiency and low costs. Comparing and contrasting methods give these professionals a picture of which may be the best option for a company.

The profile

Nisha owns her own consultancy firm, which specialises in business efficiency. During her time working in the sales team of a top accountancy firm, she began to realise that most companies unknowingly waste a good deal of time and money every day. Today, she leads a small team of efficiency experts into multi-billion pound companies and spends weeks pouring over their sales figures, employee details, business structures, budgets and customer care practices. Based on her findings, she makes suggestions for increased efficiency and profitability.

GIVE IT A TRY!

Nisha has been called into a small catering company that is looking to cut their per-event costs, but also hoping to move over to entirely organic food, which will be more expensive. However, they haven't kept very good records of their costs to date. The only thing she knows are the figures from the table below.

Cost Category	% of Budget Allotted	% of Budget Used
Office Employees	10	12
Event Staff	15	20
Chef Salaries	10	9
Food/Drink	35	29
Kitchen Tools/ Maintenance	7	10
Transportation	5	3
Office Utilities	5	6
Office Supplies	3	6
Misc.	10	5
TOTAL	100	100

1 What sort of information should she recommend the catering company gather before changing to organic produce?

2 The current budget figures would appear to indicate an overspend in all aspects of the office support – what could be the reasons for this? What may be the downside of reducing office costs?

3 Event staff costing is 5% above budget prediction. How could the company use information about previous events to plan for future ones?

Did you know that BUSINESS was so dependent upon STATISTICS?

How are Statistics used by …

… Sports recruiters?

The job

Did you ever wonder who negotiated that deal for the new striker on your favourite football team? Behind those trades, deals and new teammates are professional agents, scouts, and managers. These strategists work to source players for clubs and teams using their knowledge of the industry, the players, the club and the world of sport.

You could be:
- An agent
- A scout
- A manager
- A coach

The maths

Sports recruiters, athletic agents and even team coaches use maths every day. Player and team statistics are valuable resources for recruiters to make suitable club/player pairings. These stats are analysed, compared, deconstructed, used to forecast and serve to rank each player within a pool of candidates.

The profile

Chris has a BSc in Sports Science. He began coaching a local football team when he graduated from university. Under his guidance, the team went from only 12 players to over 25 and within two years he had led them to a regional championship. That is when Chris began to consider a career in recruitment. Today, Chris is the leading international scout for a major premiership team.

GIVE IT A TRY!

Chris is charged with filling an urgent position on a premiership football team. The coaches and managers have asked for an attacking player who can also fit into midfield and create chances. The table opposite shows three premiership players available to purchase in the

January transfer window along with their cost and performance statistics from the previous season. Use the data to produce graphs and calculations that Chris can present to the manager in order for him to make an informed choice. After the presentation, the manager buys one of the players. Which player did Chris recommend the manager buy and why?

Players	Fernando	Joe	Jamie
Goals	19	16	23
Assists	10	16	12
Appearances out of 38 possible	36	25	37
Cost in Millions £	30	18	21
Tackles made	50	200	150
Yellow cards	5	2	0
Red cards	3	1	0
Pass completion %	72	89	80

The manager is impressed with Chris' recommendations and asks him to do a detailed analysis of two of his current strikers' performances over several years because he is thinking of selling one of them. Do a statistical analysis of both players and produce a presentation.

Wayne

Year	Age	Apps	Goals	Assists
2006/7	21	35	14	20
2007/8	22	27	12	13
2008/9	23	30	12	9
2009/10	24	32	26	11
2010/11	25	16	4	10

Didier

Didier				
Year	Age	Apps	Goals	Assists
2006/7	28	36	20	3
2007/8	29	19	8	8
2008/9	30	24	5	5
2009/10	31	32	29	12
2010/11	32	24	10	10

Did you know that SPORT was so dependent upon STATISTICS?

How are Statistics used by ...

... Intelligence agencies?

The job

There is a constant stream of information or 'intelligence' to be considered where economic and national security are concerned. Those professionals who are charged with finding, understanding, interpreting and assessing secret information are called Intelligence Analysts. They handle the biggest threats to the UK and provide important support to the armed forces.

You could be:
- An intelligence officer
- An intelligence analyst
- An operational officer
- A language specialist
- A security consultant

The maths

Intelligence information does not always reach an analyst's desk in an organised and straight-forward way they must clearly assemble all the facts so they can analyse it efficiently and precisely. Analysts constantly use probability, approximations, sampling, etc. to use what little they know and determine the likely implications.

The profile

Armin is an intelligence officer. His specialty is signals intelligence, where he helps decode and understand messages intercepted from all over the world. With an MA in Modern Languages, Armin never thought he was much of a maths person. However, his fluency in several languages is a vital resource for his work. Since joining the intelligence world, Armin has travelled all over the globe and been involved in projects directly responsible for counteracting terrorism and organised crime within the UK and abroad.

GIVE IT A TRY!

1. **a** Armin is investigating complaints that a new satellite phone is not working in temperatures of above 50 °C. He knows that there are 20 areas where the phones are being used with approximately 200 phones in each location. Describe how he could determine a sample of phones to test whether or not the complaints are valid.

   b If Armin finds that 5% of phones fail in these conditions, find the probability that if ten phones are used to make calls, no more than three of them fail.

Did you know that SECURITY was so dependent upon STATISTICS?

How are Statistics used by ...

... Forensic scientists?

The job

Solving a crime requires a diverse team of experts. Forensic professionals are a vital part of the process, using their expertise in analysing evidence and looking for patterns within the clues. With every new set of evidence, crime scene investigation teams and various forensic consultants must organise and interpret the facts to determine the guilty party.

You could be:
- A forensic engineer
- A forensic biologist
- A forensic anthropologist
- A forensic psychologist

The maths

Statistics provide a gateway to proof and understanding for forensic scientists. Analysis of the evidence at hand requires everything from simple (yet very important) measurements to trigonometry (for blood spatter analysis). Probability and proportions are important tools used in forensics to support findings and to better understand the scene of the crime.

The profile

Clare is an Assistant Forensic Anthropologist working as a consultant to Hampshire County police departments. She worked as a volunteer police officer while she pursued her MSc in Human Osteology (the study of bones) and always knew she wanted to work in Forensics. Clare's primary tasks include gathering the necessary data

and samples directly from a crime scene for analysis in the lab. She and her supervisor specialise in determining skeletal age and cause of death.

GIVE IT A TRY!

A crime scene investigator on the scene of a recent bank robbery, put together this table (below) of the bullet holes discovered in the wall. You are charged with determining the likely height order of the three suspected robbers. What piece of data will you use to organise the data more clearly for this purpose? What other factors may have affected the bullet impact locations?

Bullet No.	Height of Entry	Bullet Type
1	125cm	B
2	130cm	B
3	90cm	C
4	167cm	A
5	175cm	A
6	88cm	C
7	70cm	C
8	118cm	B
9	190cm	A
10	200cm	A
11	111cm	B
12	68cm	C
13	202cm	A

1. What sort of information would you recommend the crime scene investigator collect about the bullet entry holes?

2. There were 14 other people in the bank when it was robbed. What questions should the crime scene investigator ask in order to help determine the robbers' heights?

Did you know that SOLVING CRIME was so dependent upon STATISTICS

How are Statistics used by ...

... Marketing professionals?

The job

A career in Marketing requires a good deal of creativity. Creating a campaign for a product can involve anything from major television advertisements to a free pen at the till of your local supermarket. It is not always about selling, however. Marketers promote ideas, causes and services, too.

You could be:

- An advertising executive
- A market researcher
- A public relations officer
- An online and digital marketing director
- A brand manager
- An event marketing specialist

Marketers can work in a variety of businesses. Charities, high street retailers, global banks, publishers, technology companies and arts organisations are amongst the many types of businesses that employ marketing professionals.

The maths

Statistics are the gold dust of the marketing industry. Buying tendencies, market trends, sales figures, cultural habits and demographic information determine how, why and where to promote products. These numbers can also be used to forecast the success of a potential product. Marketers use lots of techniques to gather this data, most importantly through questionnaires, online surveys and focus groups.

The profile

Daniel works in a digital marketing department on a new line of smartphone. Part of the pre-launch research he must do is to work out who his potential users will be and then determine what features they want in a smartphone. He has done a series of online surveys through social networking sites and he has also gone onto the campuses of several universities to do face-to-face questionnaires with students. He never thought he would enjoy compiling and analysing data, but he loves to see first hand the trends going on among his peers.

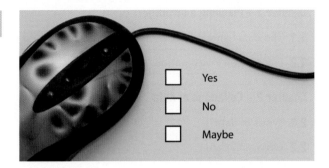

GIVE IT A TRY!

Daniel received a wide variety of responses to some market research he did last year for a new line of 'cat food that can also feed your fish!' Plot the answers on a graph from the table below to help remove some of the anomalies within the set of data.

	Non-Dog Pets	People in household
Amanda	6	3
Clare	2	2
Hannah	17	4
Andrew	0	7
Chris	4	2
Nigel	1	2
Nasim	5	4
Beverly	4	4
Sam	5	1
Nicola	11	1
Rachel	2	3

1 How could he focus his questions more specifically to better answer the question?

2 Why might it be a good idea to find out more information about the people who are seen as anomalies on the graph *before* removing them from this data set?

3 Daniel realises that some of the market research questions need improving. Write a question to find out if people are prepared to feed their cat(s) and fish the same type of food.

Did you know that MARKETING was so dependent upon STATISTICS?

Mapping chart

William Collins' dream of knowledge for all began with the publication of his first book in 1819. A self-educated mill worker, he not only enriched millions of lives, but also founded a flourishing publishing house. Today, staying true to this spirit, Collins books are packed with inspiration, innovation and practical expertise. They place you at the centre of a world of possibility and give you exactly what you need to explore it.

Collins. Freedom to teach.

Published by Collins
An imprint of HarperCollinsPublishers
77–85 Fulham Palace Road
Hammersmith
London
W6 8JB

© HarperCollinsPublishers Limited 2011

10 9 8 7 6 5 4 3 2 1

ISBN-13 978-0-00-741013-2

Greg Byrd, Fiona Mapp, Claire Powis and Bob Wordsworth assert their moral rights to be identified as the authors of this work.

British Library Cataloguing in Publication Data
A Catalogue record for this publication is available from the British Library.

Commissioned by Katie Sergeant
Project managed by Emma Braithwaite
Edited by Brian Asbury, Joan Miller and
Marie Taylor
'Real-life Statistics' written by Greg Byrd, Rob Ellis, Grace Glendinning, Claire Powis and Jayne Roper
Answers checked and proofread by Wearset
Design and typesetting by Wearset
Concept design by Nigel Jordan
Illustrations by Ann Paganuzzi
Cover design by Angela English
Production by Kerry Howie
Printed and bound by Gráficas Estella, España

Acknowledgements
The publishers wish to thank the following for permission to reproduce photographs. Every effort has been made to trace copyright holders and to obtain their permission for the use of copyright material. The publishers will gladly receive any information enabling them to rectify any error or omission at the first opportunity.

p.120 Five tablet hand held computers ©shutterstruck.com/iQoncept; p.121 Boss talking to his secretary ©shutterstock.com/alexsalo images; p.122 London, UK August 19 Kagisho Dikgacoi and Brett Holman compete for the ball playing in the international football friendly match between Australia and South Africa held at Loftus Road London 19/08/2008 ©shutterstock.com/sportsphotographer.eu; p.123 Helicopter take-off ©shutterstock.com/Wessel du Plooy; p.124 Woman working with a microscope in Laboratory ©shutterstock.com/Kurhan; p.125 Online customer survey ©shutterstock.com/Lasse Kristensen.

Browse the complete Collins catalogue at:
www.collinseducation.com